Green Energy and Technology

Climate change, environmental impact and the limited natural resources urge scientific research and novel technical solutions. The monograph series Green Energy and Technology serves as a publishing platform for scientific and technological approaches to “green”—i.e. environmentally friendly and sustainable—technologies. While a focus lies on energy and power supply, it also covers “green” solutions in industrial engineering and engineering design. Green Energy and Technology addresses researchers, advanced students, technical consultants as well as decision makers in industries and politics. Hence, the level of presentation spans from instructional to highly technical.

Indexed in Scopus.

More information about this series at http://www.springer.com/series/8059

Giovanni Campeol
Editor

Strategic Environmental Assessment and Urban Planning

Methodological Reflections and Case Studies

Springer

Editor
Giovanni Campeol
Università Iuav di Venezia
Venezia, Italy

ISSN 1865-3529 ISSN 1865-3537 (electronic)
Green Energy and Technology
ISBN 978-3-030-46179-9 ISBN 978-3-030-46180-5 (eBook)
https://doi.org/10.1007/978-3-030-46180-5

This Springer imprint is published by the registered company Springer Nature Switzerland AG
The registered company address is: Gewerbestrasse 11, 6330 Cham, Switzerland

Foreword

The Urban Governance of the Seaside Tourism: A Complex Exercise

The Municipal Administration is developing a new conception of Jesolo territory, whose strategic goals are the valorization of the environment, landscape renewal, and the completion of urban services and infrastructures. All this is aimed to offer the city a high quality of life that shall take into account the needs of residents, also considering the craft, agricultural, commercial, and accommodation activities of the territory.

The strategic view of a new Jesolo is fully expressed in the planning tool of the Land Use Plan (PAT), which is not limited to transpose the current municipal planning framework or to integrate it with higher level planning; rather, it outlines the development strategic choices. In this way, it is the fundamental moment of the planning action that the Municipality of Jesolo will then develop thanks to the subsequent Intervention Plans (IP).

Through the PAT, Jesolo is broadening its horizons with the aim of territorial integration by redesigning the urban contexts of its territory and the connection between the Lido and the historic center, focusing on the urban quality of the coastal area and the whole accommodation system.

Therefore, the PAT responds to the requests of the Municipal Administration not by promoting a general sustainable development but looking for a general quality of transformations based on specific territorial vocations.

This can be developed by defining the local environmental invariants and those arising from the higher level territorial planning while taking into account the local community's needs, ensuring the safety of inhabited centers and of the territory from the risks of hydrogeological instability. It is a planning tool that shall meet the urban needs of residents and tourists over the next 15 years.

In this context, the elaboration of the Environmental Report produced by the SEA procedure has given an important contribution to the definition of the PAT strategy, also through the drawing up of specialized studies, e.g., on the landscape interpretation and on the use of Venice northern Lagoon located in the municipal territory.

It should be emphasized that the environmental analyses were developed before the elaboration of the planning tool, and this has permitted to consider the environmental issue as a central element in the definition of the strategic goals of the new Urban Plan. In fact, the objective was to create a virtuous process between the conception of the Urban Plan and its environmental evaluation, i.e., the task for carrying out the preliminary environmental analyses, followed by the elaboration of the PAT. In this way, the environmental potentials and criticalities of the municipal territory have become essential elements of the "Plan design".

Moreover, it should be remembered that the Municipality of Jesolo has long considered the "quality" of urban and architectural planning as an essential point for defining its territorial identity. In this sense, starting from the elaboration of Kenzo Tange's Master Plan in the late 90s, the Municipal Administration has identified some emblematic questions for the PAT strategy, which have been addressed with priority.

First of all, it is clear that in the past the planning efforts to direct and regulate the growth of Jesolo were often incoherent and sometimes insufficient to effectively organize and control the city development, especially for the huge fluctuation of the population in the different periods of the year. The big effort of this administration has, therefore, been focused on defining a strategic vision for the development of Jesolo, in which the planning actions are coordinated and integrated.

The territorial competition with other resorts of the Adriatic Sea has been largely considered, with the awareness that the future of the tourist industry in Jesolo should attain the highest quality of the tourist offer for being attractive. This should not only concern the accommodation services but also the urban image, which comes from the presence of the formal quality of public places and private architectures, from the creation of new urban centralities and, finally, from the connection of different urban areas (especially the coast with agricultural areas and the lagoon).

The search for the unitary identity of Jesolo has become a strategic goal that has not been interpreted through the homologation of the various geographic scopes (lagoon, rural areas, urban areas, and coasts), but rather through the search for strong physical and functional connections of the same scopes.

Beyond the beach, the great natural attractions of Jesolo should be strongly used and suitably equipped since they are potential tourist areas, complementary to the beach and at the same time able to extend the tourism season. These goals are appropriately reflected in the PAT. For example, the lagoon scope is valorized and devised to make it become one of the attractions able to intercept various types of tourists; in summer, as a complementary offer to the beach, and in the other seasons, as a very appealing privileged factor.

Jesolo is a complex territory that must find the way to integrate seaside tourism with related sectors such as crafts and industries. In fact, the PAT supports the controlled setting up of the production units in inland planned and equipped areas, with a view to support the tourism industry and increase the number of permanent residents.

The purpose of this book is to analyze the planning experience implemented with the PAT for the Municipality of Jesolo, in which the role of the environmental assessment has acquired a strong theoretical and methodological–applicational cultural importance. This is the reason why this publication is divided into two sections: the former develops theoretical/methodological considerations on the Strategic Environmental Assessment (SEA) and the latter presents the PAT and SEA experiences in Jesolo.

Venice Province

Valerio Zoggia
Mayor of Jesolo

Contents

Strategic Environmental Assessment: An Overview of the European Experiences

Marta Bottero and Giulio Mondini

Abstract Since the last decades Strategic Environmental Assessment (SEA) has been recognized as a very important and rapidly growing area of research and application in the domain of sustainable development and spatial planning. The objective of this chapter is to provide an overview of the SEA, exploring its implementation across the different European countries and focusing on the different sectors of application. Starting from a deep state-of-the-art review, the article also highlights open questions and research needs, which allow to identify possible future recommendations.

Keywords Decision-making · Evaluation · Sustainability · Projects/plans/programmes

1 Introduction

Strategic Environmental Assessment (SEA) is a policy instrument which has been developed since the '80s as a means to influence strategic decision-making in policies, plans or programs of public bodies or responsible authorities (Partidario 2000).

More formally, SEA can be defined as a systemic process designed to assess the environmental consequences of policies, plans and programs (PPPs), in order to ensure that these consequences are included to all intents and adequately addressed from the early stages of the decision-making process, to the same level of economic and social considerations (Bottero et al. 2014; Ferretti 2012). There are many other definitions of SEA (Noble and Nwanekezie 2017) but they are all variant on this theme and they are based on the same basic principles that can be summarized as follows (Therivel 2004):

M. Bottero (✉) · G. Mondini
Department of Regional and Urban Studies and Planning DIST, Politecnico Di Torino, Turin, Italy
e-mail: marta.bottero@polito.it

G. Mondini
e-mail: giulio.mondini@polito.it

G. Campeol (ed.), *Strategic Environmental Assessment and Urban Planning*, Green Energy and Technology, https://doi.org/10.1007/978-3-030-46180-5_1

1. SEA is a tool for improving the strategic action;
2. SEA should promote participation of all the stakeholders;
3. SEA should focus on key environmental and sustainability constrains;
4. SEA should help in identifying the best option among a set of alternative solutions;
5. SEA aims at minimizing negative impacts, optimizing positive ones and compensating for the loss of valuable features and elements;
6. SEA should ensure that actions under examination do not cause irreversible impacts and damages.

The SEA procedure has been introduced in the Member States of European Union by the European Directive 2001/42/CE, on the assessment of the effects of certain plans and programs on the environment (European Commission 2001). The SEA Directive was conceived with the intention of incorporating more regard to sustainability principles into public plans and programmes across a wide range of sectors, such as transport, energy, regional development, spatial use, and so on. Indeed, its aim is to encourage the development of better plans, considering and evaluating long-term effects and wider impacts on the environment, also combining the benefits with an informed decision-making process that enables to guarantee a greater public understanding and acceptance (Nilsson and Dalkmann 2001).

Actually, after sever applications in the different Member States and many different studies, SEA is considered as a means of control of the sustainability of all the programmed actions, as a moment of comparison between objectives that may conflict with the environmental ones, and as an instrument of participation and transparency, to support an informed decision-making process.

Starting from a review of the practical experiences of SEA in the European context, this article reflects on the current state of the procedure, focusing on the main applications and the related effectiveness and exploring future possible perspectives (Fundingsland Tetlow and Hanusch 2012; Stoeglehner and Wegerer 2006; Wegerer and Stoeglehner 2004).

2 The SEA Directive

2.1 Principles and Phases of the Evaluation

The European Directive 42/2001 on the assessment of the effects of certain plans and programmes on the environment went into force in June 27th 2001, as a result of a long scientific, cultural and institutional journey. The Directive, which is composed by 20 recitals, 15 articles and two annexes, sets out the objective, the general process and the protocol and clarifies the framework that the Member States should carrying out for implementing SEA in the legal norms, whether in a separate SEA-law, or integrating it in existing laws and planning procedures.

According to the Directive, SEA develops itself in subsequent steps that consider: Screening (Is SEA necessary for the PPP under investigation?), Scoping (What are the environmental objectives of PPP? Which issues should be discussed in the assessment? Which assessment method is feasible with the available data?), Environmental Assessment (How significant are the impacts? How can these be reduced if necessary? How should these be monitored after decision-making?), Review (Is the report user-friendly and unbiased? Are all the relevant issues, including alternatives, discussed? Are the forecasts and the associated methods presented clearly?), Implementation and Monitoring (Is it clear how the transport infrastructure plan is to be implemented? Are proposals for monitoring set down clearly? Is there a mechanism for correcting any unacceptable aspects of implementation?), Consultation and Participation (Is there any plan for public participation? Is there a procedure to interact with the authorities of another country in case of transboundary corridor?), Decision (Is the SEA integrated into the planning process? Is the SEA linked with other types of assessment? Is the SEA fully considered in decision-making?) (Vincente and Partidario 2006).

Figure 1 gives a schematic representation of the overall SEA process, highlighting the participatory and integrative approach advocated by the procedure.

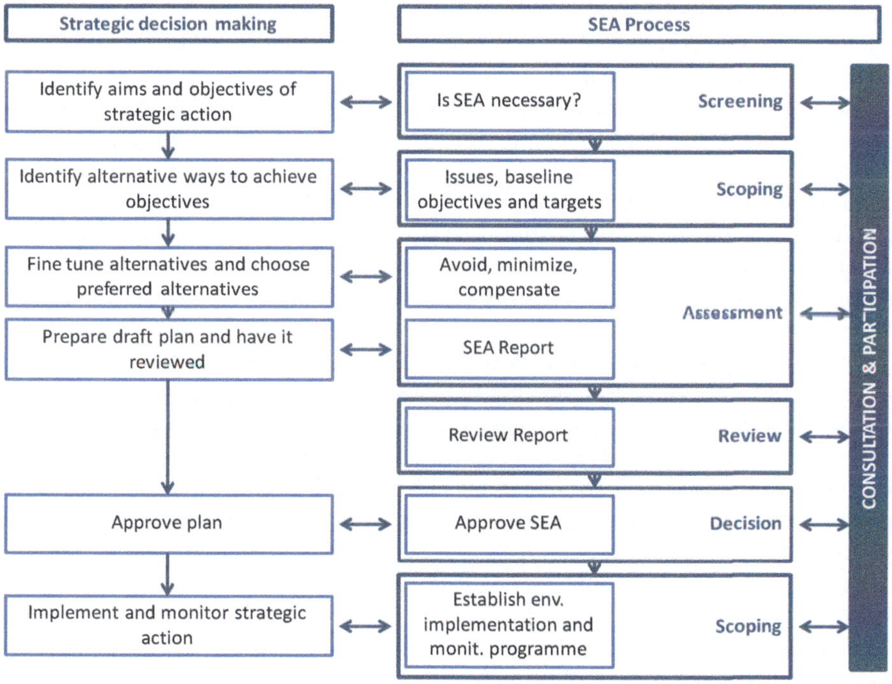

Fig. 1 Representation of the SEA process as defined by the EU Directive 42/2001

2.2 *Implementation of the Directive*

Since its adoption, the implementation of the SEA Directive has required Member States to re-think traditional approaches to planning, strengthening the consideration of long term sustainable development in planning procedures. As far as its implementation is considered, it is possible to recall the SEA Directive was receipted in different way by the Member States, so legislative framework varies between them, depending on their administrative structure. Table 1 reports an overview of national legislation in the domain of SEA in the European context. As it possible to see, the majority of the Member States have introduced the SEA legislation into their existing arrangements, e.g. Environmental Code or other legal acts; a number of Member States have also implemented several amendments to existing sectoral legislation, with specific reference to spatial planning procedures.

3 SEA state of play

As already seen, there is no universal approach to SEA and the Member States consider specific legislation frameworks and guide lines. However, it seems to be interesting to compare the situation in the different countries in order to provide a complete picture of the situation across Europe.

According to the information gathered by a review of different studies and European documents (European Commission 2016; Weilinad 2010; Dusik and Sadler 2004), it is possible to define the average number of concluded SEA procedures carried out each year, for the period 2007–2014. The number of SEAs performed per year has a wide range of variation, that goes from 2 procedures per year in Malta to 3000 procedures per year in Germany (Fig. 2). This variance strictly depends on the size of the State and also on its SEA implementation and legislative framework. Despite these large differences, it is possible to observe that small Member States carry out few SEAs per year, as in the case of Luxembourg with 15 SEAs for year. It is also worthy notice that Belgium has only 9 SEAs per year, which could reflect the national government's lack of competences in this field. With particular reference to the Italian situation, the information collected in a specific study carried on by the European Commission (2016) highlights that the number of completed SEAs in the period 2009–2013 is 600 on average per year with 1000 screening procedures per year.

As far as the sectors of application are concerned, Fig. 3 summarizes the general level of diffusion of SEA across the different countries. As it is possible to see, the domain of spatial planning, including town and country plans, covers the majority of SEA completed applications. Also water plans play an important role in the existing applications, followed by transport and energy plans.

Other interesting findings emerge from the overall analysis of the SEA procedures in Europe in the same period are reported in Fig. 4. Examining this graph

Table 1 Type of national legislation transposing the SEA Directive

Integrated legislation				
	Specific SEA legislation	Environnemental code/Environnemental Protection Act	EIA legislation	Sectoral legislation
Austria	X[a]			X[b]
Belgium Federal	X			
Belgium—Flanders Region	X		X	X
Belgium—Brussels Capital Region	X			X
Belgium—Wallonia Region		X		X
Bulgaria	X	X		
Croatia	X	X		
Cyprus	X			
Czech Republic			X	X
Denmark	X			
Estonia			X	X
Finland	X			X
France		X		
Germany	X		X	X
Greece	X			
Hungary	X	X		
Ireland	X			X
Italy	X	X[c]	X[d]	
Latvia	X		X	
Lithuania	X			X
Luxembourg	X			X
Malta	X			
Netherlands		X	X	X
Poland		X	X	X
Portugal	X			
Romania	X			
Slovakia			X	
Slovenia		X		
Spain	X		X	
Sweden		X	X	
UK	X			X

[a] Some Provinces
[b] Federal level and some Provinces
[c] SEA/EIA at National/Regional level
[d] Some Regions
Source elaboration from European Commission (2016)

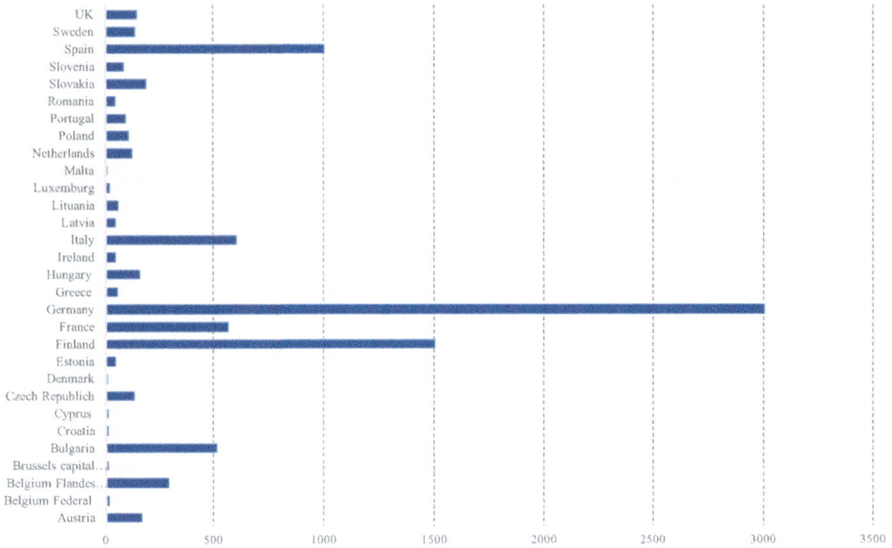

Fig. 2 Average number of SEA procedure carried out each year 2007–2014 (*Source* Elaboration from European Commission 2016)

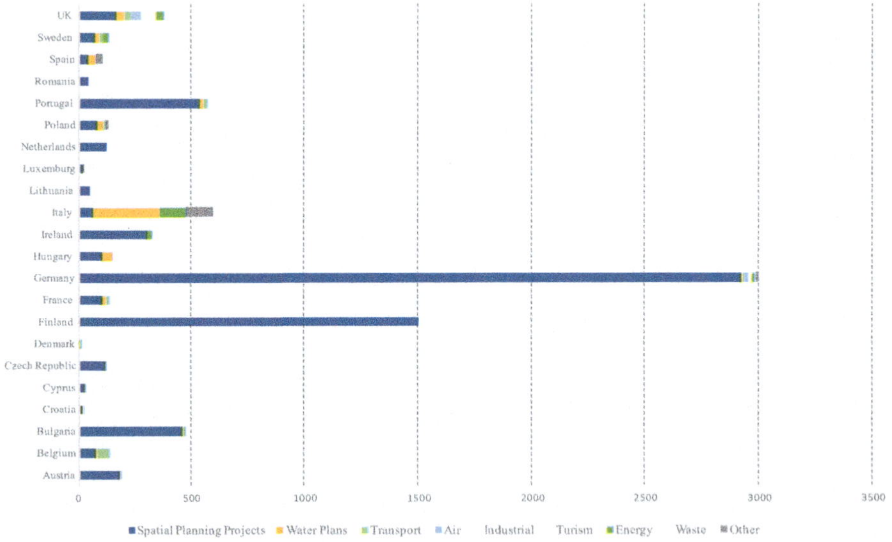

Fig. 3 Most common types of plan and programs considered in the SEA procedures in the period 2007–2013 across the different European countries (*Source* Elaboration from European Commission 2016)

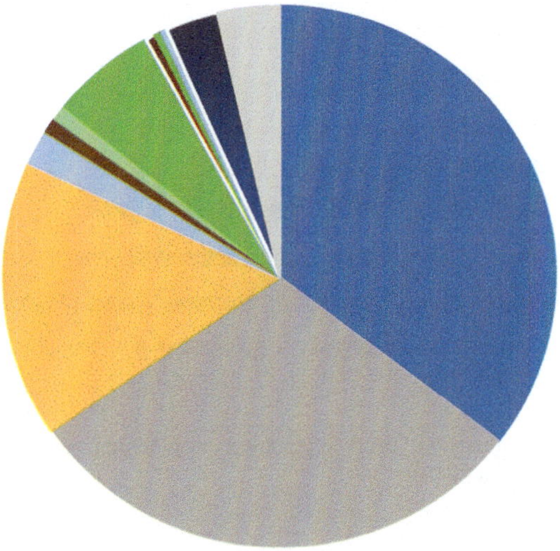

Fig. 4 Distribution of the SEAs procedures in the different sectors of application in Europe in the period 2007–2013 (*Source* Elaboration from European Commission 2016)

it is confirmed that the areas of prevailing applications are those related to spatial planning projects, even if other categories are important such as forestry, cultural heritage, industrial activities and tourism. As far as these categories are considered, it is important to define that Italy, despite being a small country, is one of the major countries that is experimenting the SEA in these minor areas of application, undertaking around the 20% of the total SEAs procedures in the cultural heritage and tourism domain.

4 Discussion and Conclusions

After having explored the SEA applications in the European context, it is interesting to reflect about its effectiveness and performance. This is a very complex task because the SEA success strictly depends on the context where it is applied and from the specific legislation adopted by each country. In this sense, many scholars and researchers have made an attempt of measuring SEA effectiveness using rigorous analytical tools (van Doren et al. 2012; Sheate and Eales 2016; Partidario and Fisher 2004; Fischer 2007). Generally speaking, these studies distinguish three main components of the concept, that can be described as follows:

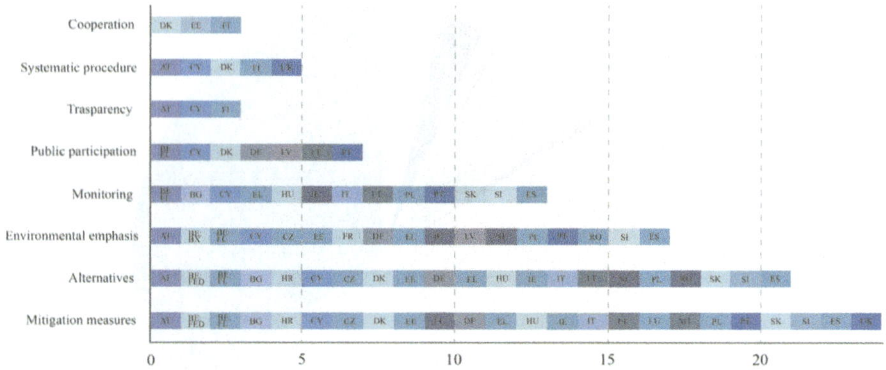

Fig. 5 Perceived influence of SEA applications in planning processes (*Source* Elaboration from European Commission 2016)

- substantive effectiveness, which allows to understand if SEA achieves its purpose and is able to inform the decision-making process;
- procedural effectiveness, which looks at whether SEA follows the legal requirements;
- transformative effectiveness, that takes into account more intangible benefits, such as the creation of social learning processes and knowledge exchanges.

In a recent survey carried on by the European Commission (2016), a questionnaire was submitted to the representatives of the different Member States in order to collect information about the results of SEA implementation. As far as the SEA influence on planning processes is concerned, the study highlights that the most significant consequences are related to the increase of the environmental emphasis at the preliminary phases of the process, to the consideration of a number of alternatives and to inclusion of mitigation measures (Fig. 5).

These findings are also confirmed by the relevant literature in the domain of SEA where the most relevant benefits are connected to the following elements (Marsden and De Mulder 2005; Sadler and Verheem 1996; Jones 2005; Theophilou et al. 2010):

- SEA allows for a solid decision-making process, allowing transparency in the procedure, supporting the participation of the stakeholders and anticipating threats and negative impacts;
- SEA permits to create and develop a credible alternative as a result of its ability to lead and shape projects;
- SEA ensure strategic thinking, appraising actions and scenarios since the very early stages of the decision process.

However, different authors put in evidence also limits of SEA, focusing on the insufficient attention devoted to monitoring activities in the completed procedures which lack in the use of specific indicators and indexes able to measure the real effectiveness of the considered plans. Another weakness is related to the difficulties in the

Table 2 SWOT analysis resulting from the SEA state-of-the-art-review

Strengths	Weaknesses
• SEA increasingly leads to changes in PPP contents • SEA is becoming more integrated into planning processes • Raising awareness around environmental implications of decisions • Leading to more transparent processes • Widespread application • Contributing to capacity building	• Capacity of SEA to exert limited influence • Long-standing shortcomings related to SEA process limitations
Opportunities	**Threats**
• Better integration with decision-making • Facing global challenges: considering environmental limits, ecosystems services and climate change issues • Strategic SEA supporting good governance	• Unclear role and aim of SEA • Overload of SEA expectations • Recession causing limited resources available for SEA

Source Elaboration from Fundingsland Tetlow and Hanusch (2012)

correct definition of compensatory measures for balancing the negative environmental impacts of the plans. In this sense, a very promising innovation is related to the introduction of ecosystems services within the SEA practice that could support the quantitative evaluation of the alteration of the environmental system. A final drawback emerging from the analysis of SEA reports and scientific documents is related to the poor culture of the practitioners in the field of environmental assessment who tend to perceive SEA as an exercise to meet legislative requirements, rather than a process which adds real value to the planning process.

In line with the aforementioned considerations, Fundingsland Tetlow and Hanusch (2012) summarise the main strengths, weaknesses, opportunities and threats reflected during the state-of-the-art review of the SEA procedure (Table 2).

In conclusion it is possible to say that even if SEA is still evolving and has not yet reach its full potential, nevertheless there are important arguments supporting its positive role in raising awareness of the environmental implications of strategic decisions (Mondini 2016, 2019).

References

Bottero M, Ferretti V, Mondini G, Pomarico S (2014) Enabling public participation in strategic environmental assessment: an application of multicriteria analysis. Strateg Behav Environ 4(2):99–130

Dusik J, Sadler B (2004) Reforming strategic environmental assessment system: lesson from Central and Eastern Europe. Impact Assess Proj Appraisal 22:89–97

European Commission (2001) Directive 2001/42/EC of the European Parliament and of the Council on the Assessment of the Effects of Certain Plans and Programmes on the Environment. Brussels

European Commission (2016) Study concerning the preparation of the report on the application and effectiveness of the SEA Directive (Directive 2001/42/EC), Luxembourg: Publications Office of the European Union, 2016, ISBN 978-92-79-60785-1. https://doi.org/10.2779/725510
Ferretti V (2012) Verso una valutazione integrata di scenari strategici in ambito spaziale. I modelli MC-SDSS. Celid, Torino
Fischer T (2007) Theory and practice of strategic environmental assessment: towards a more systematic approach. Earthscan, London
Fundingsland Tetlow M., Hanusch M (2012) Strategic environmental assessment: the state of the art. Impact Assess Proj Appraisal 30(1):15–14
Jones C (2005) Strategic environmental assessment and land use planning, an international evaluation. Earthscan, London
Marsden S, De Mulder J (2005) Strategic environmental assessment and sustainability in Europe—How bright is the future. Reciel 14(1)
Mondini G (2016) Integrated assessment for the management of new social challenges. Valori e Valutazioni 17:15–17
Mondini G (2019) Sustainability assessment: From brundtland report to sustainable development goals. Valori e Valutazioni 23:129–137
Nilsson M, Dalkmann H (2001) Decision making and strategic environmental assessment. J Environ Assess Policy Manage 3(3):305–327
Noble B, Nwanekezie H (2017) Conceptualizing strategic environmental assessment: principles, approaches and research directions. Environ Impact Assess Rev 62:165–173
Partidario MR (2000) Elements of an SEA framework: improving the added-value of SEA. Environ Impact Assess Rev 20(6):647–663
Partidário MR, Fischer TB (2004) Follow-up in current SEA understanding. In: Morrison-Saunders A, Arts J (eds) Assessing impact, handbook of EIA and SEA follow-up. Earthscan, London, pp 224–247
Sadler B, Verheem R (1996) Strategic environmental assessment: status, challenges and future directions. Ministry of Housing, Spatial Planning and the Environment, The Hague
Sheate WR, Eales RP (2016) Effectiveness of European national SEA systems: how are they making a difference? In: Sadler B, Dusik J (eds) European and international experiences of strategic environmental assessment: recent progress and future prospect
Stoeglehner G, Wegerer G (2006) The SEA-Directive and the SEA-Protocol adopted to spatial planning—similarities and differences. Environ Impact Assess Rev 26:586–599
Theophilou V, Bond A, Cashmore M (2010) Application of the SEA directive to EU structural funds: perspectives on effectiveness. Environ Impact Assess Rev 30:136–144
Thérivel R (2004) Strategic environmental assessment in action. Earthscan Publications Ltd., London, UK
van Doren D, Driessen PPJ, Schijf B, Runhaar HAC (2012) Evaluating the substantive effectiveness of SEA: towards a better understanding. Environ Impact Assess Rev 38:120–130
Vicente G, Partidario MR (2006) SEA—enhancing communication for better environmental decisions. Environ Impact Assess Rev 26:696–706
Wegerer G, Stoeglehner G (2004) Implementing Strategic Environmental Assessment (SEA) into spatial planning—an ex-post survey of spatial development concepts of Lower Austria. In: SUSTAIN Life—Secure Survival II. Conference Proceedings Prague, September 22–25. ISBN: 80-213-1197-5
Weilinad U (2010) Strategic environmental assessment in Germany—practice and open questions. Environ Impact Assess Rev 30:211–217

The Strategic Environmental Assessment in Italy

Paola Andreolini and Giuseppe Bonavita

Abstract The application of the SEA has led to a more structured set of plans and programs, as objective oriented tools and more related to sustainability decision-making models. In the relationship between the different types of plans and the hierarchical scale of the assessment tools, the political and regulatory needs still prevail rather than the needs of the territory and the environment. Actually, this evidence is limiting the effectiveness of participation practices. In Italy, the support of the Ministry of the Environment (MATTM) on the state level of planning and also on regional level of plans and programs, as main contributor in SEA consultations, has the aim of enhancing the diffusion of best practices and checking the implementation status of environmental policies in the Country.

Keywords Strategic environmental assessment · Sustainability goals · Planning · Participation

1 Introduction

Thirteen years have passed since the transposition of Directive 2001/42/EC by Italian law. Important results have been achieved, others, more ambitious, need further regulatory and operational efforts. The execution of the environmental assessment at the beginning of the decision-making process has revealed its importance for the expression of the potential of designing that the SEA has within the planning activities.

The attention of the legislator is being addressed on the monitoring and follow-up of the assessment procedures, which at local level could be subject to distortions. It

P. Andreolini
Core Component of Analysis and Evaluation of Policies (NUVAP), Department of Cohesion Policies, Italian Presidency of the Council of Ministers, Rome, Italy

G. Bonavita (✉)
Technical assistance for permitting and EU's regulation to the Italian Ministry of the Environment, Rome, Italy
e-mail: bonavita.giuseppe@minambiente.it

G. Campeol (ed.), *Strategic Environmental Assessment and Urban Planning*, Green Energy and Technology, https://doi.org/10.1007/978-3-030-46180-5_2

is equally important to strengthen the awareness actions of the community, through measures that refer to the promotion of increasingly active forms of participation aimed at the formation of consensus and public debate.

Starting from the study of data on procedures and reporting activities carried out by the Ministry of the Environment, is possible to obtain important items that can be considered in a process of SEA refitting and for implementing guidelines to promote and share with the local proposers.

1.1 The Support to the Assessment Processes: The Role of the Ministry of the Environment

In these last years, within an heterogeneous framework for the elaboration processes of environmental policies, the relation between decision-making and planning has been developing into a technical, social and cultural process coming from the interaction of new various factors. The application of Directive 42/EC, while keeping the same principles since 2001, has in fact favored a gradual paradigm shift in the processes of environmental assessment. The transposition of the SEA regulations in the EU's Member States laws, with the needs to integrate the evaluation process into complex administrative procedures, has been almost exclusively oriented towards the proceduralization of the different evaluation stages, considering the qualification of the cognitive, decisional and participatory process as the essential condition to effectively address and interpret the many needs that should be considered in the elaboration of plans and programs.

The Public Administration, responsible for decision-making, is promoter and manager of these processes; therefore, with a view to self-improvement, the monitoring of the correct execution of the current assessment processes is an essential activity that shall verify the effectiveness of the regulations and guidelines to which the tools of the territory management and environmental protection relate.

The periodical review of the national implementation of the SEA[1] procedures gives a detailed picture that highlights various and recurring critical fields; these have not been solved yet and the reasons are to be found especially in the complex mechanisms that regulate the functioning of the Public Administration, and in the realtionship between administrative needs and technical appraisal.

The Italian Ministry of the Environment, Land and Sea (MATTM), when implementing the principles of the EU's regulations, enhance the systematic application of the SEA in the elaboration process of the programming tools for the governance of the territory, under regional and local competence. The Ministry promote the integration of the goals of environmental sustainability, the implementation of new effective know-hows and assessment techniques, as well as the sharing of environmental information in the participatory processes.

---

[1]The Report on the implementation of the SEA is published annually by the Directorate-General of Environmental Assessments of the Ministry of the Environment.

At national level, the MATTM plays the role of Competent Authority for issuing the SEA "expressed opinion", in collaboration with the Ministry of Cultural Heritage and Activities, and for the SEA screening opinion; moreover, it gives support and collaboration to the different stages of the evaluation process.

Thanks to its environmental expertise, it is also consulted within a lower level SEA procedures, mostly at regional level. That level corresponds in many cases to the first programming and planning step of the important tools for the territory management and development. As a result, local planning significantly influences the implementation of national environmental policies, effectively contributing to achieve sustainability goals, with special reference to the mitigation and adaptation to climate changes, to the reduction of pollution, and sustainable energy management; Italy is internationally committed to pursue this goal with its environmental and sectorial policies.

In this context, the Ministry promotes, through the tools made available by the European Funds, the strengthening of the administrative capacity in terms of the procedural and technical management of the environmental procedures, by supporting the regional and local governments in the exercise of their functions and powers. (See Par. 3: The "door-to-door" SEA Service: meeting the regional and local administrators, and the citizens).

The comments sent within the framework of the SEA consultations, concerning the first level of regional territorial and sectorial programming and planning, published on the portal va.minambiente.it, aim at giving a feedback on technical and methodological contribution to the realization of the assessment activities and to the construction of the Plan sustainability. The consultations mainly concern the regional sectorial plans and, cyclically, the operational programs for the use of SIE funds; less frequently, they are addressed to the plans of national parks. However, the Ministry is not involved in the SEA consultations on local urban plans; in this case, it contributes only indirectly through methodological, legislative and fact-finding guidelines and observations conveyed through the contributions supplied within the framework of higher-level planning tools. In 2016, 17 contributions were sent for as many consultations (Table 1).

The analysis of the procedures under consultation is focused on general aspects or else on the principles of the environmental action and of the SEA. The comments formulated analyze in detail the specific aspects where it seems necessary to specify

Table 1 Consultations on SEA of plans/programs in which the ministry took part in 2016

Year	Number of consultations	Regions/public Authorities	Sector of P/P under consultations
2016	17	Abruzzo (2), Campania, Calabria (2), Lazio (2), Apulia, Marche, Friuli Venezia-Giulia (3), Basilicata, Molise (2), Tuscany (2)	Energy (2), Parks (3), Forests, Waste management (6), Air quality (3), Landscape, Mining activities

and integrate information that could modify the assessment framework. Less often, some comments and suggestions are proposed on the planning and programming activities, especially where it is estimated that are available some other options and solutions that might ensure a better achievement of the sustainability goals defined by Community and sectorial regulations.

1.2 Implementation of the SEA at National and Regional Levels

From the annual reports, it is possible to obtain many valid elements for surveys and in-depth studies on the real implementation of the SEA and on the ability to significantly impact on planning and programming at different levels. While considering that every year a variable quantity of data on the activities of local authorities cannot be surveyed, with the subsequent underestimation of the total number of national procedures, the territorial coverage is however significant and it has permitted to elaborate also specific in-depth studies, especially at the level of local authorities.

The SEA procedures and the screening of urban Plans concerning local level represent the most surveyed procedures in 2016 (80% of all SEAs and 94% of screenings); their management often requires regulations among the most controversial of Part II of Legislative Decree 152/2006 concerning the SEA: the modifications of "small areas" and the environmental assessments of implementing tools for urban Plans (Art. 5 par. 8, supplementing Art. 16 of Act No. 1150/42) have an important impacts on the sustainability of territorial transformations and administrative disputes.

The value measured in 2017 (procedures concluded in 2016) continued to be on the same levels of the previous year, highlighting an activity of ordinary updating of the planning tools that was no longer linked with the initial rapid increase determined by the need to adapt the local Plans to the SEA laws. Moreover, on a programming level there are no relevant spikes, since the tools already approved entered into force in 2014 and will be valid until 2020.

The following data show an important heterogeneity of the situations detected in the different Regions and Autonomous Provinces (Tables 2 and 3).

The widespread application of the Screening SEA procedure in local planning did not allow a complete recognition of data. Almost all types of Plans or Programs subject to this procedure are inter-municipal, municipal and implementing urban local Plans. In terms of the values obtained, the numerous SEA screenings are of crucial importance. These procedures are started in relation to the approval of zoning plan variations or of unassessed local implementing Plans, or also as variations of planning tools already approved. These tools are attributed to a large numbers of minor modifications and have significative environmental effects (Table 4).

The screening procedure has not only a much shorter duration than the SEA but it is also evidently less complex. The screening is in fact a self-standing procedure, less constrained by the approval process of the Plan/Program, and it is therefore

Table 2 Non-state SEA procedures concluded from 2011 to 2016

Region/Autonomous Province	2011	2012	2013	2014	2015	2016
Piedmont	33	1	1	1	5	31
Valle D'Aosta	0	2	1	5	2	1
Lombardy	215	261	319	229	63	61
Trento	1	21	40	25	44	10
Bolzano	0	3	2	5	2	5
Veneto	79	0	21	24	32	52
Friuli Venetia Giulia	3	7	13	5	9	4
Liguria	1	1	n.a.	11	7	8
Emilia-Romagna	121	104	112	130	89	106
Tuscany	21	20	47	64	33	27
Umbria	2	2	1	4	6	1
Marche	10	10	12	10	5	5
Lazio	3	n.a.	0	2	5	4
Abruzzi	0	n.a.	n.a.	n.a.	2	1
Molise	n.a.	3	1	3	5	0
Campania	18	20	n.a.	8	6	9
Apulia	9	19	6	11	20	17
Basilicata	0	2	3	3	2	4
Calabria	11	0	2	2	0	6
Sicily	n.a.	n.a.	2	6	n.a.	4
Sardinia	10	9	7	8	36	11
Total	**537**	**485**	**590**	**556**	**373**	**367**

Table 3 Non-state SEA procedures concluded from 2011 to 2016 divided by types of plans/programs

Types of plan or program	2011	2012	2013	2014	2015	2016
ESIF plans and programs	7	4	1	23	20	1
Regional, provincial and urban territorial plans	15	15	18	20	18	18
Sectorial and management plans	38	39	49	55	73	54
Inter-municipal, municipal and implementation urban plans	477	427	520	447	262	298
Total	**537**	**485**	**590**	**556**	**373**	**367**

Table 4 SEA screening procedures in Italy from 2011 to 2015

Competence	2011	2012	2013	2014	2015	2016
State	0	0	0	8	4	3
Non-state	953	1184	1060	1120	1034	1217

Source MoE—Report 2017

Table 5 Non-state SEA screening procedures in Italy from 2012 to 2016 and outcomes

Result/year	2012	2013	2014	2015	2016
Subject to SEA	51	67	67	83	61
Not subject without prescription	348	559	617	496	616
Not subject with prescription	781	427	406	455	546
Total	**1180**	**1053**	**1090**	**1034**	**1217**

subject to the influence of fewer variables; this condition substantially simplifies the management of procedures, with a consequent reduction of procedural times. As concern the data collected by the MATTM, almost the whole the SEA Screening of Plans and Programs under regional competence clearly excludes the SEA, with or without prescriptions and recommendations referred to their implementation.

Overall, as shown by the below table, about 95% of the screenings conclude on the exclusion from the SEA; this means that, according to regulations, will not be carried out further environmental analyses (Table 5).

1.3 The Assessment of Urban Plans

The Plans under local authorities' competence represent the most important part of the procedures subject to the SEA and the SEA screening. The Plans here considered include urban and land-use Plans, Plans on transport, development and recovery, strategy, trade and use, as well as partial or complete variants and updates of these Plans. Instead, the urban Plans of provinces and metropolitan areas are not considered.

One can observe how in this type of tools the strategic environmental assessment is prevalently addressed to the evaluation of implementation Plans, when these are not completely defined in the Plans (and therefore subject to screening), or in those cases where they modify municipal planning. To a lesser extent, the periodical updates of municipal planning tools are subject to assessment (Table 6).

As regards the data on the screenings concluded in 2016, it is underlined how this procedure is almost exclusively dedicated to the screening of local urban planning

Table 6 Non-state SEA screening procedures in Italy from 2015 to 2016

Survey year	Total SEA under regional/local competence	SEA of inter-municipal, municipal and implementation urban plans	SEA percentage of urban Plans on total procedures (%)
2015	373	262	**70**
2016	367	298	**81**

Source MATTM—Report 2017

Table 7 SEA screening procedures in Italy from 2015 to 2016

Survey year	Total screening under regional/local competence	Screening of inter-municipal, municipal and implementation urban plans	Screening percentage of urban plans on total procedures (%)
2015	1034	993	**96**
2016	1217	1144	**94**

tools, with special reference to the modifications called "minor" or that involve small areas of the territory planned in the tools already subject to the SEA (Table 7).

As highlighted in the abovementioned report, about 95% of the screenings conclude on the exclusion of the SEA, and this percentage still slightly increases if it takes account of the screenings carried out on the local urban Plans, thus reaching almost 97%. In relation to this statistical evidence of data, some regulatory integrations have been introduced over time, with the aim to define more clearly the impact of the principles of assessment non-duplication and of the environmental non-significance of the effects caused by the Plans or Programs, regarding especially urban Plans (Art. 5 Act No. 10 of 2011). In fact, for their complexity and in relation to the need to intercept the changing requirements of the territory, the urban Plans represent the tools more likely to undergo changes or partial modifications.

1.4 Issues and Innovations Within the SEA Implementing Framework

According to the data provided by the survey and the supporting activities carried out by the Ministry in collaboration with regional authorities, it is confirmed that there is a problem of application for the regulations governing the realization of the SEA in the detailed planning, especially in interpreting the definition of "small areas" and "minor modifications" at the screening stage of the urban implementing tools.

Controversial effects also arise from the application of Art. 6 par. 12 of Legislative Decree 152/2006, which excludes the SEA from any modification to urban Plans on the basis of decisions authorizing single works that have, by law, effect of variants. The application of this last article has often been subject to court proceedings, which means to the interpretation by competent authorities.

In relation to the efficiency of the procedures, the trend of devolving the functions of Competent Authorities to local bodies, although shared by many in compliance with the principle of subsidiarity, involves undeniable difficulties connected with the organizational and financial weakness in the administrative structures of the demographically smaller municipalities, especially if the territory is particularly broad or it is characterized by environmental susceptibility. However, the most serious criticalities are related to environmental monitoring: the findings obtained at regional and local levels are poor and are not published by local authorities. Therefore, it is plausible to consider that the obligations concerning ex-post monitoring are in many cases disregarded. The absence of monitoring activities precludes any possible confrontation and verification of the SEA findings, therefore limiting the possibility of any further evolution of the procedure; consequently, the whole planning and SEA activity is limited within the scope of predictions.

The non-implementation of monitoring plans is strongly penalizing any possible innovations or updates of Plans and Programs in relation to the adaptation to the environmental effects that can be seen through the monitoring output; therefore, also the opportunity to timely avoid any possible unexpected impacts is penalized. Despite important criticalities, there are some positive trends mainly concerning the improvement of the transparency and publicity for such measures. The best innovations concern the use of computer applications for data management and the use of the web for the dissemination of information, especially through the implementation of web-GIS systems.

Regulatory innovations are mainly focused on procedural aspects, through experiences of integrated SEA/EIA procedures or screenings integrated with the SEA preliminary stage.

2 The Building of the Sustainability in Planning

The SEA application has favored an approach of Plans and Programs that seem increasingly to comply with the need to clearly explicit measurable strategic and environmental goals and related operational actions, "the Plan goals and actions, first established unconditionally and—so to speak—intuitively, are now selected in relation to their environmental compatibility, in connection with the principles of sustainability, following procedures that are not completely typified but are gaining ground on an experimental basis".[2]

The adoption of a complementary logical structure, between planning and assessment models, has not only brought undoubted practical facilitations, mainly arising from the opportunity to have a specific organization of all information in a useful way and ready to be processed and evaluated, but also, from a strictly methodological viewpoint, it has led to a progressive objectivation of these models, both in literature and in the elaboration of processes (cultural and scientific).

Although there are no legislative references prescribing the use of specific analyses and assessment techniques, in a view of procedure standardization a crucial contribution has been given by guideline documents laid down, among the others, by the Ministry of Environment and the ISPRA, and realized at local and regional levels where most SEA procedures are developed in Italy.

[2]Colombo et al. (2012), *Dalla pianificazione urbanistica alla pianificazione ambientale*, in "Pianificazione urbanistica e valutazione ambientale". Le Penseur, Brienza (PZ), p. 74.

The elaboration of thematic contributions, the realization of in-depth study workshops and on the job support activities have decisively helped to develop “networks” between the administrative authorities involved in the procedures with the aim to share good practices and to form local experts and professionals on the environment.

Many issues found in the SEA procedures, especially in the sections showing the inefficiency of the procedure, can be attributed to inconsistencies or shortcomings in the instruments under evaluation and also to aspects unequivocally related to the planning and programming quality. Moreover, this criticality is manifest in the long-lasting misalignment of the Plan elaboration times and the assessment process, so that the latter is ineffective.

When starting the planning activities, the definition of some methodological goals could be considered as the precondition for an integrated evaluation of the Plan/Program, with special attention to:

- Evaluability of the plan strategy, insofar as the sequence (general goals—specific goals—actions/measures) is clearly formulated and detailed;
- Flexibility and feasibility of the planning and programming projections where the numerous territorial and environmental variables, that could define different reference and development scenarios, have been adequately considered;
- Correct involvement of the territory government in the mosaic of the planning and programming tools of the same level and in relation to the higher-level or lower-level hierarchy, meeting the need of not creating “gaps” in planning and evaluation or, rather, of not overlapping the regulatory structure on the competence of other tools;
- Of a framework building for the environmental sustainability of the national/regional plan to be used for identifying the most suitable environmental goals for the Plan, that shall be developed following the priorities and criticalities, also at local level.

2.1 Items and Circularity of the Process: Needs, Goals, and Intervention Strategies

Although “stages” are generally referred to for outlining the development of the SEA process, this simplification should not lead one to think that consequential stages will be closed once they have been developed; it is rather an “open” circular process in which the main elements—needs, goals and actions, together with their measurement through the different types of indicators (context, realization and contribution)—are developed and analyzed as the planning and programming process gradually develops.

The building of the sustainability of a Plan or a Program starts from the analysis of the context and from the identification of the criticalities and needs in relation to territorial, social and economic peculiarities. The Plan can therefore elaborate the most adequate intervention strategy that is liable to affect and give a proper "contribution" to achieve the objectives of environmental sustainability and to overcome and/or correct the ongoing criticalities.

The main role that the Plan should perform to obtain an efficient evaluation lies in the correct way to collect the necessary information in order to build a comprehensive strategic framework and to identify and use the logical links between criticalities/goals/actions. This means to develop a conceptual model that may foresee the characterization of an early scenario, the definition of a target scenario, and the operations to be carried out in order to obtain the expected results.

However, what environmental goals shall integrate the Plan goals and what intervention strategy shall be adopted?

The identification of the environmental components or issues that the Plan can affect is closely connected with the type of the Plan: sectorial plans, such as those for the waste or water resource management and for transport, or for territory governance, will repeatedly affect certain environmental scopes, e.g. air and water quality, soil consumption, noise, waste management and landscape.

The peculiarities of the contexts—presence of sensitive areas (areas subject to major incident hazards, protected natural areas or sites of the Natura 2000 Network, coastal or inland territories) or of critical areas (e.g. those where pollution limits set by laws are being exceeded)—shall integrate the reference environmental aspects.

The choice of the environmental goals (SEA priority) with which integrating any plan objectives without a direct environmental goal, shall be made by identifying the same goals as those required by the reference strategic framework laid down in the Plan or at regional/national levels.

On 22 December 2017, the Inter-ministerial Committee for Economic Programming (CIPE) approved the National Sustainable Development Strategy (NSDS)[3] that is based on the wide reference framework of global sustainable development and represents the first step to implement the principles and goals of the Agenda (2030) on a national and regional level.[4]

The linkage with the national and regional Sustainable Development Strategies realized after the transposition of the SEA Directive into the national law, with Art. 34 of Legislative Decree 152/2006, channels the contributions of individual Plans to the implementation of the priorities defined for each reference scale towards an efficient contextualization of the choices made.

Thanks to the informative and participatory processes that define the regional strategies, the coherence with the National Sustainable Development Strategy is realized following the peculiarities of the territories, and is therefore developed on the basis of real priorities and needs resulting from the participation of entire communities.

---

[3]Previously Italy had already adopted a National Sustainable Development Strategy for the environment, approved by the CIPE on 2 August 2002.

[4]The Agenda (2030), adopted in 2015 by the United Nations with 192 signatory States including Italy, identifies 17 goals (Sustainable Development Goals-SDGs) and 169 universal targets, that involve all countries and are based on the integration between of the three dimensions of sustainable development (environmental, social and economic).

This reference strategic tool for sustainability, from which to select the associated goals and targets, enables to address at different planning levels the contributions of individual Plans in a view to achieve the objectives set in higher-level and in national Plans through a targeted implementation process that becomes increasingly detailed and accurate in the transition to a new scale.

Once the territorial needs (in terms of criticalities and opportunities to enhance) and the Plan objectives (integrated with the goals of environmental sustainability) have been identified, the Plan may elaborate its intervention strategy by defining specific objectives, actions, and local interventions.

The intervention strategy shall be identified after the elaboration of "many alternative strategies". These should be chosen among strategies considered more "environmentally-friendly", through the comparison of intervention alternatives compared with the ongoing reference trend scenario, i.e. scenario determined by the evolving context in its most representative environmental components, covered by the Plan, as if the Plan is not implemented.

The concept of "carrying capacity", i.e. the capacity of absorbing further effects caused by identified actions on one or more components or specific areas, entails the need to establish reference targets, or "thresholds", within which to assess the impact increase and, above all, the indicators fit for measuring it. Moreover, starting from the assumption that no plans or programs are free of effects, this concept enables assessing the absorption level, or the "carrying capacity", that the plan's reference scope can accept without significant variations. Therefore, it is possible to detect and quantify the set of the most appropriate actions for limiting the impacts of the plan and for the achievement the targeted environmental objectives.

The process is outlined in its main elements (needs, goals, actions) in the first "stage", that corresponds to the so-called "scoping" and to the interaction with environmental experts. It is helpful for defining the "sphere of influence or extent of the information" that shall be included in the environmental report and shall determine the "environmental dimension of the Plan or Program" (stage of the Plan guidelines). When the process is further developed, it becomes fundamental for the assessment in the elaboration of the Environmental Report; in this stage, the analysis becomes more detailed as the choices of the plan (actions, localizations, times) are identified.

Two further elements extend transversally to the whole developing process:

- The construction of a common knowledge basis (environmental data and information, indicators);
- Participation as an information instrument and choice-sharing.

For developing the assessment process, it is necessary to measure it, quantitatively and qualitatively, in the characterization of the context state, in the identification of the targets and the estimation of their attainment. It shall evaluate the identified alternatives and, finally, provide for monitoring.

This aspect decisively impacts on the evaluation quality, and is nevertheless still largely jeopardized by the current difficulties due to the absence and, in many cases, the fragmentation of data, the gaps in the certification of the data sources, the unavailability of complete or homogeneous time series, the absence of indicators that can be sufficiently disaggregated.

Operationally the integration of the environmental assessments, carried out in the different scales of the planning tools and among the same environmental assessments (SEA-EIA-AIES) on a basis of common knowledge, as well as the setting up of reference frameworks for objectives-indicators at regional level would make it possible to use homogeneous data on the local planning scale, also as regards the realization of aggregate frameworks (e.g. for the tools and assessment processes with the associations of municipalities).

2.2 The Participation in Decision-Making

The other cross-cutting aspects of the SEA process on participation deserves separate investigation.

Although the model suggested by the European Directive has not solved the question of the responsibilities yet, since the transposition of the regulations into the national law is left to the Member States, the Directive provisions on the separation between competent and proceeding authorities seem to be understood on a purely functional level.

Instead, the attention of the EU legislator is being transferred on actions aimed to devolve responsibility to the community, through measures directly referred to the process transparency and to the promotion of increasingly active participatory forms for forging agreements and public debate,[5] through measures that have already been established within the SEA.

The issue of participation, in a context of integrated planning, is a critical point in all the experiences considered. At legislative level, the matter is analyzed especially in relation to the need to promote forms of inter-sectorial participation, among the different planning levels and territory management and among the different types of actors. In most cases, this type of participation finds its niche in the process, in specific institutional moments (preliminary consultations, planning conferences, inter-institutional meetings, etc.) already laid down by rules, on the basis of an existing decision-making structure in which the stakeholders are required to give their opinions or to suggest integrations.

---

[5]Legislative Decree 18 April 2016, n. 50 Italian Public Contract Code Art. 22 (Transparency in stakeholder involvement and public debate), implementing Directive 2014/24/EU on public procurement, has introduced the formula of public debate. On the same line, the integration to Legislative Decree 152/2006, following the entry into force of Legislative Decree n. 104 of 2017, that provides for the public enquiry during SEA consultations: *"pursuant Art. 24, par. 3, first sentence, the competent authority may require that public consultation shall be held in the form of public enquiry, with charges payable by the proposer, within a maximum time limit of ninety days. The enquiry shall be concluded with a report on the work carried out and a judgment on the outcomes, accomplished by the competent authority"*.

In (2017) the Ministry of the Environment, at the end of a long period of studies and diffusion of techniques and good practices for the participatory processes started with the 2007/2013 programming, published the "Linee guida sulla Sintesi non tecnica del Rapporto Ambientale" (Guidelines on the non-technical Summary of the Environmental Report), in which the complexity of information and of the issues in the documents on public participation was identified as part of the problem. However, in the light of the procedures and case studies analyzed, there are still numerous and wider perplexities on the methods of managing the consultation stages in relation to the interaction with the planning process rather than to the SEA.

The question is complex and it is subject to various experimentations, more or less efficient, especially in some Regions called "virtuous" and in big cities. However, the question that shall be analyzed concerns the method for making the participation process more effective and profitable in a framework of integrated planning, that may therefore involve the stakeholders and promote collective instances through concentration mechanisms and equal agreements between the parties.

Currently the process is structured, al least in its goals, for directing the participation so that institutional decisions may be socially accepted, assuming that the goals of the Plan and of the SEA have been defined before consultations. In this context, the sustainability goals, being transposed into sectorial directives, are generally defined even before those of the Plan.

Essentially the decision-making step remains hidden to the transparent procedure of the formulation and evaluation of the plans, therefore neither declared nor clearly analyzed in the framework of what should represent the third actor of planning, i.e. the participation process.

As regards the relationship with stakeholders and the public, the SEA has a much wider meaning than the environmental sustainability, since it has a strategic character affecting all aspects of life quality.

In order to meet these needs, established long ago in the Arhus Convention, the SEA procedure should be considered as a tool first for raising awareness, so that those involved and interested may help to contextualize the real goals, even though they do not define them.

Avoiding the simple adoption of a "consensus" model would represent part of the solution to the problem of the marginal role of participation; on the other hand, a model for achieving a scenario in which the SEA defines the decision-making framework would seem too demanding.

The basis for consensus and the need for advertising in the very first stages of the decision-making process, highlighted by recent literature and environmental authorities, and strongly supported by the EU—which funded and promoted the new and numerous procedures (Agenda 21, Programming activities for the ERDF and ESF funds)—may contribute to qualify the public as a bearer of ideas and innovations, thus leading to the production of shared, accepted and participatory instruments of territorial governance. All this is possible when aiming to implement a process of collective learning, followed by a real phase of participation in the territorial decisions.

It should be remembered that most SEA procedures[6] are developed at local level; the aspect of the participation stringently recalls the responsibility of public authorities. The participation processes, legally mandatory, should be somehow calibrated following the planning levels they are referred to, and should actually evolve towards a concertation process as a necessary element for ensuring the feasibility of the plans. Consequently, also in terms of participation, it is necessary to underline the assumption of accountability of the local authorities, which shall guarantee effective participation processes by educating and informing the citizens.

This last aspect represents the most important move for the development and cultural growth that the SEA can make for the community, but at the same time it is often disregarded despite the formal respect of the times, advertising and consultation procedures.

3 The "Door-to-Door" Sea Service: Meeting the Regional and Local Administrators as Well as the Citizens

The objectives to ensure sustainable development finds in the environmental assessment processes, especially those within the SEA, the most efficient tools for reaching the general goals of national and regional environmental sustainability and ensuring an integrated and comprehensive protection of the environment in all its components as well as the connection and coordination between environmental and sectorial policies at different institutional levels.

For reaching these targets it is necessary to implement a new organization of the decision-making model and to realize a structured support able to reinforce the specific evaluation skills inside the administrations, with different roles and competences, involved in the planning and evaluation processes, in a view to improve the management capacity of the co-decision processes typical of the SEA.

The individual Plans and Programs shall in fact ensure, each in its own scope, the coherence between the actions and individual interventions realized and the real and concrete targets to be reached.

Many aspects concerning the environmental assessment processes still need be detailed, from the participation and consultation to the peculiar aspects of the environmental assessment (elaboration of environmental and programming frameworks, definition of reference scenarios, generation and evaluation of alternatives, construction of systems for environmental monitoring, identification of feedback mechanisms after monitoring activities).

[6]In the Report on the SEA state of implementation—2017, published by the Ministry of the Environment, concerning the procedures completed on the national territory in 2016, the SEAs of urban (municipal) plans accounted for more than 90% of the total surveyed.

Generally, for all environmental assessment processes, the poor quality of available environmental data and the impossibility for the Administrations, in terms of planning and research centers, to have a shared system of environmental data and certified information tools are still among the main critical points affecting the quality of environmental reports and also of the evaluators' opinions.

Moreover, the low availability of appropriate leading tools for the definition of scenarios, the identification of the targets of reducing environmental impacts, and the recognition and selection of the most effective measures to realize them, together with poor planning capacity, frequently jeopardize the success of interventions.

Some shortcomings, such as the absence of "a project" for public participation and of appropriate facilitation instruments may render relatively ineffective the consultations of both public and environmental experts.

These aspects, especially for provincial and local administrative authorities with delegated powers for the SEA and EIA, are added to fragile situations due to the shortage of staff and specialized technical expertise, which are necessary conditions for the elaboration of the evaluation processes.

The current legislative framework regarding the SEA introduces new needs for institutional cooperation by recognizing to the national and regional environmental authorities a new and more incisive role in the process—where the integration between the government levels shall be stronger—as well as new management procedures and organizational models that need common criteria and methodologies (central government and regional authorities) for an efficient and homogeneous application of the legislation. The simplification process started by the legislation calls for integrated procedures between the different assessments, SEA, EIA and AIES (horizontal integrations) and within the SEA processes of the planning tools at different levels (vertical integration) (Fig. 1).

For an effective execution of these processes, it is necessary to strengthen the specific competencies in the administrations involved by providing them with a special support in terms of tools, know-how and information, therefore facilitating informed and structured participation.

The National Operational Programs "Governance and Technical Assistance" (NOP GAT 2007–2013) and "Governance and System Actions" (NOP GAS 2007–2013) make it possible to conduct cooperative activities and actions whose goal, for the Lines of action in the field of Environmental Assessments, is to strengthen the governance capacity of the regional and local public administrations of the Convergence Objective (Calabria, Campania, Puglia e Sicily) in the cycle of 2007–2013 joint programming, for implementing and managing the environmental assessment processes. Another goal is to achieve the targets of sustainable development at the heart of programs, and of integration of the environmental concerns in the definition and implementation of the sectorial policies established in the different Plans and Programs.

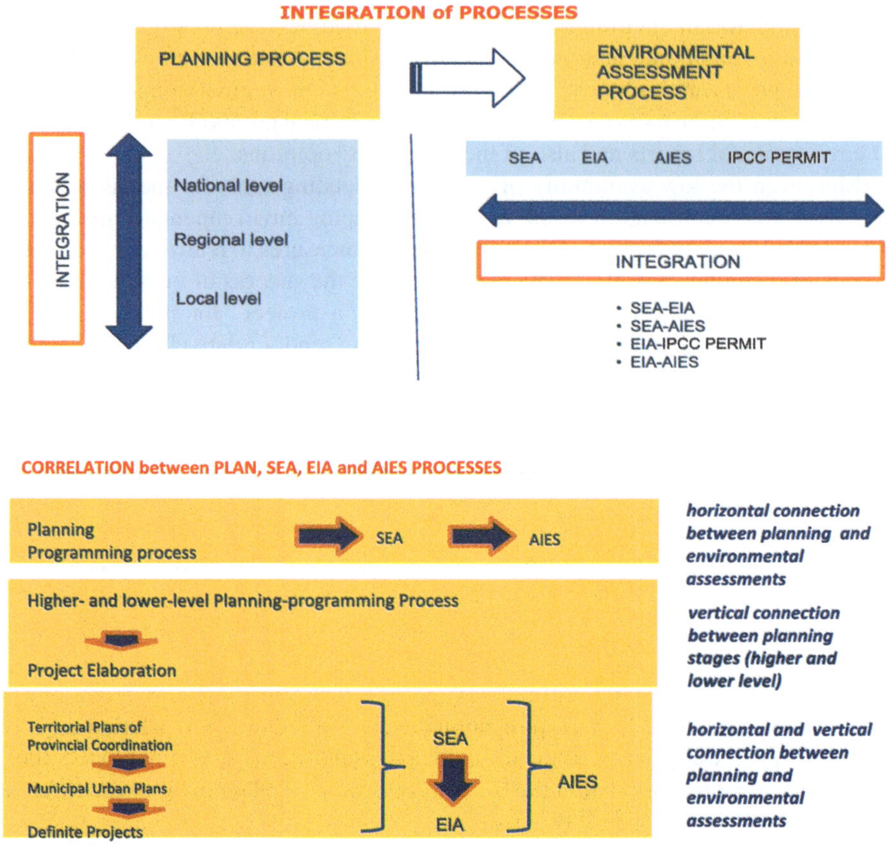

Fig. 1 Levels of planning integration—assessment and connection between assessment tools

Beyond providing the staff and experts with a methodological and policy-making support in a view to implement the Strategic Environmental Assessment and the Environmental Impact Assessment (Environmental reference Frameworks, identification of objectives and reference targets for the different levels of planning, technical contents, operational guidelines, thematic in-depth analyses through sectorial studies, realization of training activities and "on the job" mentoring), the experience conducted has started an efficient "network" system for studying, confronting and "replicating" the technical and methodological support tools, also through the application of case studies, in the least developed regions.

Unione europea
Fondo sociale europeo

Ministero del Lavoro e delle Politiche Sociali
Direzione Generale per le Politiche per l'Orientamento e la Formazione

Governo Italiano

per il tuo futuro
Programmi operativi nazionali per la formazione e l'occupazione

Operational Program NOP Governance and System Actions (ESF) 2007–2013
Intervention line 7. Sustainable development

The National Operational Program "Governance and System Actions 2007–2013" (NOP GAS) is designed to contribute to strengthen the models of environmental *governance* of the regions in the Convergence objective, through the progressive grounding in the ordinary practice of operational criteria, technical tools, management procedures and organizational models that favor the institutional innovation in the fields of environmental integration, Strategic Environmental Assessment and Environmental Impact Assessment.
The Intervention Line 7 "Sustainable development" is articulated in two Actions:

- Action 7.A—"Horizontal actions for environmental integration" under the responsibility of the DG SEC
- Action 7.B—"Support actions to the processes of Strategic Environmental Assessment (SEA) and to the procedures of the Environmental Impact Assessment (EIA)" under the responsibility of the Directorate-General for Environmental Assessments of the MATTM

Overall aim and general objectives

The main general objectives of Action 7.B can be summarized as follows:

- Strengthening the *governance* capacity of national, regional and local Public Administrations to implement the objectives of sustainable development in the Joint Programming 2007–2013 and the integration of environmental concerns in the definition and implementation of ongoing programs
- Ensuring the effective implementation of the EU and national environmental legislation on the homogenous and structured conduct of the environmental assessment processes

Specific objectives of the Project

The main specific objectives of Action 7.B are the following:

- Starting system and strengthening actions aimed to improve the *governance* capacity of the Public Administrations
- Ensuring the exchange and sharing of experiences and technical-scientific contents for carrying out the processes of the Strategic Environmental Assessment and the Environmental Impact Assessment
- Ensuring, with operational guidelines, the development of criteria and technical directions for laying out the EIS
- Developing shared and homogeneous methods for the analyses and environmental assessments

Activities and products

Conferences; workshops e seminars; thematic workshops; information and awareness raising; actions for the development of competences; exchange of expertise; elaboration of sectorial studies, methodological documents and operational guidelines; pilot projects; publications; websites; database and good practices

Among the "field" initiatives, together with those of the local administrations involved and actively participating in the planning processes and environmental assessment, it is necessary to mention the following.

In Calabria: theme workshop "Equalization aspects of Planning and Strategic Environmental Assessment of the urban Plans of Calabria Region", in which the Directors of the Regional Department for the Environment and the Territory, the coordinator and the technical staff of the MoE program meet the local administrators of all the municipalities of the Calabria provinces responsible for laying down their own planning tools and relevant SEAs. The goal is to support the elaboration, by supplying technical and methodological tools and procedural explanations, and to share the direct experiences of rapporteurs who illustrate specific study cases.

In Salerno: the workshop for thematic in-depth studies "Operational guidelines for the coordination and integration of the SEAs at provincial and municipal levels: the case of the PTCP (Territorial Plans for the Provincial Coordination) of SALERNO" has brought together the officials of the Province of Salerno and of the Municipalities in the Salerno Province that are involved in local activities of planning and environmental assessment.

During the training activity, developed in 5 days, all participants have actively exchanged opinions on the in-depth themes connected with the SEA of the PTCP of Salerno. The workshop was aimed at enhancing the aspects of procedural coordination and integration of the contents. More specifically, starting from the guidelines of the PTCP Environmental Report and in relation to the progress of the PUCs (Municipal urban plans), being defined in the same Province, it was suggested to investigate some specific aspects of the SEA process such as: "How were the plan alternatives in the SEA elaboration identified? What kind of mitigation measures were realized after the impacts recorded? What were the contents/data/indicators elaborated to integrate the Assessment of Implications and the SEA? How has the Plan for Environmental Monitoring been laid down, taking account of the entire chain of municipal and over-municipal plans involved?"

In Naples: the theme workshop "The elaboration of the Strategic Environmental Assessment for the PTCPs—The study case of the PTCP of Naples" goes backwards, with the aim to verify the quality of the process now coming to an end. Addressed to the officials of the Province and Municipalities of Naples, the workshop has involved officials of other administrations—Calabria and Umbria Regions—committed to the same themes in their own institutions, and the officials working on the PTCP of Naples. Stepping in at an advanced stage of the draft plan, the workshop analyzes some key aspects, such as rewards, equalization and compensation, SEA indicators of the PTCP, construction of the reference scenario, generation and evaluation of alternatives and of the environmental monitoring system.

In order to analyze the experiences carried out and to acquire technical tools for supporting and analyzing the main aspects of the evaluation processes, it is helpful to view the documents and products realized and published in the website pongas.minambiente.it in the section "Azione 7B", in which it is possible to download the outcomes of the workshops, the publications, the guidelines and sector studies on the different issues, the documentation and the general or specific teaching materials on all the modules held. The excellent outcome of the Program has enabled a new proposal of system actions with the aim to improve the competences of the Public Administration in key sectors for the environment and sustainability.

With the CReIAMO PA project—Competencies and networks for the environmental integration and the improvement of government organizations, funded within the NOP Governance and Institutional Capacity 2014–2020—Axis 1 "Development of the administrative and institutional capacity for the modernization of the Public Administration"—Action 1.3.3 "Interventions for improving the central and regional administrative capacity for the integration of environmental sustainability", the MoE has developed a strategy to raise the efficiency levels of the Public Administration in the environmental field, which shall involve all Regions, differently from the previous experience focused only on the Convergence Regions (Calabria, Campania, Puglia and Sicily), through the strengthening of the administrative capacity, the development of the e-government and the improvement of the multi-level governance. The project is articulated in nine lines of action, with training and on the job tutoring; it presents a specific action, LQS1—Environmental assessments—Actions for improving the efficiency of the SEA and EIA processes concerning Programs, Plans and Projects referred to the aspects of the assessment procedures. The goal of the project is to increase the administrations' capacities for implementing the environmental assessment procedures in an efficient and simplified way, by applying uniform criteria on all the national territory, thus ensuring the quality of processes and their interaction, the certainty of the time schedule, the transparency of the administrative action, the participation of the citizens, and the sharing of information on past and ongoing procedures, on the progress of works and on the outcomes of monitoring activities.

References

Colombo L et al (2012) Dalla pianificazione urbanistica alla pianificazione ambientale. In: Pianificazione urbanistica e valutazione ambientale. Le Penseur, Brienza (PZ), p 74

Ministero dell'Ambiente e della Tutela del Territorio e del Mare (2017) "Rapporto 2017 sull'attuazione della VAS in Italia", a cura di Luciana Polizzy e Giuseppe Bonavita. http://www.va.minambiente.it/it-IT/Comunicazione/DettaglioDirezione/1196

The Agenda 2030, adopted in 2015 by the United Nations with 192 signatory States including Italy, identifies 17 goals (Sustainable Development Goals-SDGs)

Scenario Analysis and Strategic Environmental Assessment

Francesca Torrieri

Abstract The European Union (EU) has introduced the Strategic Environmental Assessment (Directive 2001/42/EC), with the aim to support the implementation of actions in the long run and evaluate the impacts of policy, plan and programs The strategic thinking and the participation of the stakeholders to the decision-making process represent the main innovations in contrast to Environmental Impact Assessment (EIA). In this context the paper presents scenarios analysis developed within Future studies literature as a tool to facilitate strategic thinking in SEA and to support an integrated planning process based on the participation of the main stakeholders involved. Starting from an analysis of the literature on case studies developed in different contest where scenario methods are test within SEA process, a cognitive methodological framework is pro-posed with the aim to underline the role of scenario analysis in each phase of SEA decision-making process, from the definition of the main scenarios to the evaluation of the strategic action and the monitoring after the plan implementation.

Keywords Scenario analysis · Strategic environmental assessment · Cognitive methodological framework

1 Introduction

The EU has introduced Strategic Environmental Assessment (SEA Directive 2001/42/EC) with the aim of ensuring a high level of environmental protection and promoting the integration of environmental and social considerations in the strategic decision-making processes (article 1 of the directive). Conceived under the philosophy of environmental impact assessment (EIA) as an assessment process appropriate for policies, plans and programs (PPPs) (Wood and Djeddour 1992), SEA is today

F. Torrieri (✉)
Department of Industrial Engineering DII, Università Degli Studi Di Napoli Federico II, Naples, Italy
e-mail: frtorrie@unina.it

G. Campeol (ed.), *Strategic Environmental Assessment and Urban Planning*, Green Energy and Technology, https://doi.org/10.1007/978-3-030-46180-5_3

considered a tool to support the formulation and implementation of strategic initiatives, to evaluate the impact of the actions selected and even play a political role in decision making process (Partidário 2015; Jiliberto 2011; Bina 2007). The philosophy of SEA overcome the EIA approach in terms of facilitates strategic thinking, enabling transitions toward sustainability, and promote the participation of all the stakeholder involved (Partidário 2012). In the SEA, particularly innovative compared to the traditional planning process is the participatory decision-making process to protect legitimate interests and the creation of the conditions for consensus among stakeholders on actions to be implemented in a territory (Torrieri and Batà 2017).

SEA is about exploring desirable outcomes, determining what is needed to achieve those out-comes, and identifying and assessing the potential implications of alternative strategic initiatives (Noble and Gunn 2015). The focus is on design sustainable and resilient future alternative scenarios (Slootweg and Jones 2011), instead to extrapolate it on past trends, condition or events (Noble and Gunn 2015; Partidário 2007).

As argued by Noble and Nwanekezie (2016), SEA can be characterized as strategy-based in contrast to EIA approach more focus on evaluation of alternative predesign project and their impact, capturing more recent thinking about SEA as a process for driving institutional change. What differentiates the innovative approach of SEA are the purpose(s) of the its application and the extent to which the strategic principles, are more or less reflected in its design, intent and implementation (Noble and Nwanekezie 2016).

A Meta analysis of case study developed at European level (Ireland, Environmental protection Agency 2012) shows that generation of reasonable alternatives is one of the biggest challenges in SEA. Nevertheless, the extent and quality of how alternatives are developed and considered is limited: planners often consider that there are no reasonable options for their plan, and that consultants' alternatives are not 'reasonable'. While in best practices examples is highlight as the public consultation can stimulate creative thinking, and that the integration of SEA procedure in the planning process, can improve the capacity to identify and resolve issues at a very early stage, which should eliminate the need for changes later in the process.

In this context, future studies and scenario methods can play an important role in stimulating creative thinking and support the formulation of an open range of options. In Table 1 advantage of scenario approaches in SEA are reported. As Table 1 show, scenario approach can stimulate creative thinking moving from results determinates by the status quo and on the base of deterministic analysis to a more flexible and qualitative picture of open future driven by proactive vision.

In respect to the latter consideration the paper presents different approaches to scenario analysis with specific reference to the methods introduced in the literature in the context of SEA with the aim to propose a cognitive methodological framework within the SEA process.

The paper is organized as follow: in the first paragraph scenario analysis is presented with specific reference to the main approaches developed during the time; then a review of the literature is commented, with the aim to evaluate point of weakness and straightness of the scenario methods adopted in different context of SEA.

Table 1 Progressive advantages of scenario approaches in strategic environmental analysis

From	To
– Focus on quantified variables	– Focus on qualitative pictures
– More emphasis on details	– More emphasis on trends
– Results determined by status quo	– Results based on future images
– Deterministic analysis	– Creative thinking
– Closed future	– Open future
– Statistical-econometric tests	– Plausible reasoning
– From quantitative to qualitative	– From qualitative to quantitative
– Single track thinking	– Multi-track thinking
– Reactive problem driven	– Proactive vision driven
– Multiple implicit assumptions	– Transparent simple assumptions
– Limited set of options	– Open range of options
– Model-determined mind	– Alertness to signals of uncertainty

Source Adapted from Nijkamo et al. (1997)

On the bases of the literature review a cognitive methodological framework is proposed with the aim to understand the contribution of scenario methods in the different phases of SEA procedure. Final considerations are reported in the conclusion.

2 Scenario Analysis

Products in the traditional approach to strategic planning of an unknown future attempts to foresee all its aspects extrapolating it from the existing trends. However, in the planning field, forecasting is extremely complex since it is characterized by complex situations, whereby various groups of actors with conflicting objectives take part to the decisions process. Actually, the decision process is characterized by a high degree of uncertainty related to the future, especially in our contemporary dynamic and risky era (Beck 1999) where changes are swift and difficult to control. Uncertainty is, in fact, linked to the behavior of the actors involved, to the unexpected or undesired impacts of the decision and to exogenous risk factors.

In respect to the above consideration, the literature (Bell 1997; May 1996; Schwartz 1991; Khakee 1999) has proposed approaches and methodologies typical of Future Studies to cope with uncertainty. Within Future studies, scenario analysis includes ample ranges of methods and techniques that can be used in many fields.

From the ancient time, people have used scenarios as a tool for indirectly exploring the future of society and its institutions (von Reibnitz 1988; Wilson 1978). However, as a strategic planning tool, scenario techniques where introduced by military planning in 1950a at RAND Coorporation. After that scenario methodology was used in the 1960s for social forecasting, public policy analysis and decision-making

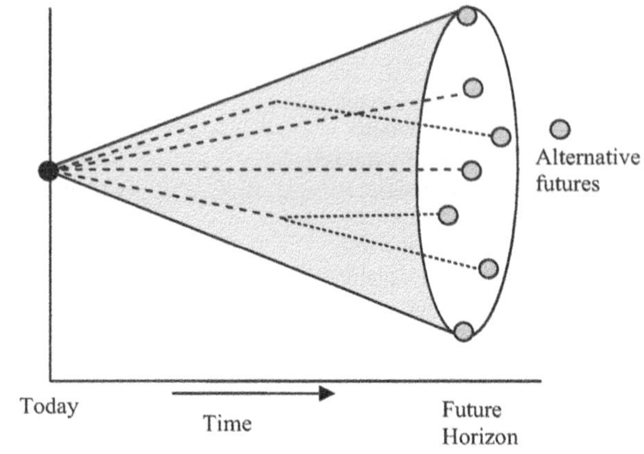

Fig. 1 Scenario development

especially in the USA[1] and France[2] (Kahn and Wiener 1967; Wack 1985). Other interesting studies where developed in the Netherlands (Van der Heijden 1996).

Scenarios essentially represent coherent and feasible 'maps' of the future. So scenario analysis can be seen as the study of the strategic behavior of actors in relation to true actions or events. Many definitions have been developed in the literature (Vleugel 2000; Chermack and Lynham 2002; Schwartz 1991; The Batelle Institute in 1996), the most meaningful studies were those of Kahn and Wiener (1967) that define a scenario as "a hypothetical sequence of events built in order to focus the attention on casual knots of decision processes".

According to the above definition, a scenario does not represent a forecast or a preferred development of an actual situation; it is instead a set of coherent and believable descriptions that represent different visions of alternative futures, described according to a chain of events (Fig. 1).

[11] In the USA the most meaningful studies were those of Kahn and Wiener (1967) who within the RAND Coorporation undertook numerous military commissions for the US armed forces. Founders of the Huston Institute developed their use in the 1960s, coining the phrase "to think the unthinkable" in relation to a forecast of the threat of a thermo-nuclear war. During the 1970s the Research Institute of Stanford proposed an innovative structured approach for scenario building under the guidance of the illustrious luminaries Willis Barman, Harnold Mitchell, Oliver Markley and Marie Spengler. Particularly important was the contribution made by the consultancy organizations societies that operated in this field: the Batelle Institute, the Global Business Network (Schwartz 1991), the Northeast Consulting, and the Future Group.

[2] Within the French school, the Centre d' Etudes Prospectives conducted fundamental studies in scenario approaches to long-term planning called 'prospective thinking' or 'You Prospective' (Berger 1967). This approach reportedly emerged as a consequence of the repeated failure of 'classical' forecasting approaches. In this context, an extremely important contribution is that of Godet (1986), whose interest was mainly in morphological analysis and in existing relationships among different actors in the process of scenario building.

In other words, scenario development can be seen as a tool to explore the future rather than to foresee it, to build contexts to support the decision, thus lowering the level of uncertainty and raising the level of knowledge. A scenario provides a context to think and reason about factors, relationships among actors and situations that answer the question "What would happen… if?". Scenario elaboration is the task of an interdisciplinary team and helps to understand the points of strength and weakness of a project.

According to the cyclical development process proposed by Kolb (1984), starting with the concrete experience, we can build models of possible futures achieved via different chains of events that then have to be valued and verified against different contextual conditions. Systematic information helps us to ensure that the future vision is built on the basis of the knowledge acquired during the process, in relation to the expectations, values, needs and events that can take place over time. The process is a cyclical-learning process in which new information is continually being turned into knowledge and verified on the base of the hypotheses formulated (Torrieri and Nijkmap 2008).

The indeterminateness of the events that shape the possible future, causes the planners, the policymakers and all the subjects involved in the process of scenario building to react in different ways in relation to an uncertain and not deterministically predictable situation; in fact, it is possible to identify different ways to face uncertainty: to ignore the uncertainty; to identify and to specify the degree of uncertainty; not to do anything and wait until the uncertainty naturally reduces; to accept the idea that a condition of uncertainty exists and to act in conscious way to manage it; or to face the uncertainty not as a threat, but as an opportunity to model the future in a creative way. This last attitude is sometimes defined in the literature referred to as the 'no-regret strategy' (see Nijkamp 1994), in the sense that defined strategies may also effective, even if the conditions are substantially modified over time. Therefore, scenarios do not claim to foresee the future, but are a toolset for describing possible chains of events that can determine alternative spatial transformations. This appears to be very useful within a Strategic Environmental Evaluation process, because they can support the definition of the alternative of action, thus reducing the degree of uncertainty set by future dimensions.

Many methods have been developed in the field of scenario analysis; scenario development can be divided in two basic approaches:

Future Backward: we depart from the individualization of possible alternative futures and investigate the models and the choices that could bring about such scenarios;

Future Forward: we depart from the analysis of the existing conditions and, on the basis of the evolution of present situation, possible futures are examined.

Departing from such generalizations, scenarios usually have four dimensions (Inayatullah 1996):

- *Status Quo*: it is assumed that the future will be a continuation of the present;

- *Collapse*: this appears when the system cannot keep on growing anymore, or when existing conditions bring it to a state of irreversible breakdown;
- *Steady State*: this is based on a return to a past condition, imagined or real;
- *Transformation*: fundamental changes are hypothesized that can be in both values and technological innovation, or they may be political and economic changes.

In terms of a methodological approach, we can identify three fundamental approaches:

- *Intuitive logic*: Global Business Network and Shell;
- *Probabilistic modify trend (Trend impact analysis and cross impact analysis)*: Future Group;
- *La Prospective*: French school.

Where probabilistic modify trend is a quantitative approach, whereas intuitive logic is more qualitative oriented; la prospective use a combination of qualitative and quantitative tools and the researcher describe it as a mix of intuitive logic and probabilistic methods (Amer et al. 2013).

Scenarios are also classified as descriptive and normative: the first are extrapolative in nature and presents a range of future likely alternative events, while the second are goal directed, respond to policy planning concerns in order to achieve desired target.

Scenarios can be also classified on the base of topic (problem specific versus policy scenario), or of the scope (one sector versus multisector scenario) or focus of action (environmental versus social or politics) and level of aggregation (micro verso macro scenario) (Mietzner and Reger 2005).

The classical Handbook of Futures Research (Fowles 1978) dedicates around 30 pages to the description of the existing methodologies: for example, Delphi techniques, game theory, brainstorming, check lists, morphological analysis, the cross-impact matrix, analyses and extrapolations of trends, regression analyses, etc. Moreover, since the publication of this book, over the years many other connected techniques have been developed, above all in the field the strategic planning and management. The most popular are: Interactive Cross Impact Simulation (INTERAX/SMIC), Interactive future simulation (IFS), Trend Impact Analysis (TIA) and Fuzzy cognitive maps.

One of the principal challenges in this field today is not so much to develop new technologies, but rather to test a process that can integrate in an efficient and effective way those already existing "… the right tool for the right job…" (Ratcliffe 2002, p. 21).

In the next paragraph a review of the methods used in the field of Strategic Environmental Evaluation are reported and commented.

3 Scenario Analysis and SEA: A Review of Case Studies

The literature review has been carried out through the electronic database Scopus. The articles have been selected based on a keyword research in which the following combination has been used: "Strategic environmental assessment" and "scenario analysis".

A total of 56 papers have been found in the last 10 years and then a further sorting has been provided according to the abstracts. 14 papers have been considered relevant and suitable for the aim of the research. For each articles selected the following information has been recorded (Table 2):

- Author(s);
- Year;
- Decision context;
- Case study;
- Type of analysis.

The analysis of the literature shows that although there is a clear need for futures studies in several tools for environmental systems analysis, it is interesting to note that the combination of scenario analysis and SEA is rather limited, in fact only 14 case studies was that selected where a consistent scenario analysis is carried out. The majority of the case studies examined are located out of Italy, especially in China.

The case studies have been classified in the following categories on the base of the main topic, the scope of work and the focus of actions:

- *Climate scenarios* referred to climate change projection and designed to represent future climate scenarios and to evaluate the impact. In Larsen et Al. strategic thinking and scenario analysis was used to reduce uncertainty in climate change within SEA process of municipal Danish planning system.
- *Socio economic scenario* where demographic trend, sensitivity and adaptability of economic system represents driving forces for scenarios development.
- *Environmental scenarios* encompass future environmental factors and conditions that consist to threats to natural ecosystem and environmental consequences of land use (Bragagnolo and Geneletti 2013; Geneletti 2013; Bai et al. 2010)
- *Water resource scenarios* that try to simulate and evaluate the impact of urban transformation on water river (Zhou et al. 2008).
- *Land use scenarios* based on urban growth and desirable planning, actor interrelation and neighborhood connection (Sizo et al. 2015; Bragagnolo and Geneletti 2013; Rozas-Vasquez et al. 2014; Petrov et al. 2011; Wang et al. 2004).

For most of the selected studies all these categories are strongly interrelated and successful environmental scenarios studies usually combined elements of the abovementioned categories. Many methodologies have been used for scenario development more oriented on quantitative analysis as Markof chain technique, Spatial analysis, stochastic simulation, social analysis based on interviewed and probabilistic analysis.

Table 2 Case studies

Author	Decision context	Title	Case study	Year	Type of analysis
Sizo A. Noble, B.; Bell, S.	A scenario-based approach to strategic environmental assessment for wetland trend analysis and land use and land cover (LUC) modeling in an urban environment	Future analysis of urban land use and wetland change in Saskatoon, Canada: an application in Strategic Environmental Assessment	Canada	2015	Markof Chain technique and Remote sensing data
Bragagnolo, C.; Geneletti, D.	Land use scenarios are generated and then assessed against a set of environmental indicators	Dealing with land use decisions in uncertain contexts: A method to support Strategic Environmental Assessment of spatial plans	Italy (Milan)	2014	Land use scenarios and Spatial analysis (GIS)
Rozas-Vasquez, D.; Pena-Cortes, F.; Geneletti, D., Rebolledo, G.	Development and evaluation of different scenarios for wetlands	Scenario modeling to support strategic environmental assessment: Application to spatial planning of coastal wetlands in la Araucanía region, Chile	Cile	2014	Spatial explicit scenarios and GIS
Lamers, M.; Liggett, D.; Tin, T.	Strategic approach to environmental governance in Antarctic	Strategic thinking for the Antarctic environment: The use of assessment tools in governance	Antartic	2013	Partecipatory scenarios

(continued)

Table 2 (continued)

Author	Decision context	Title	Case study	Year	Type of analysis
Larsen, S.V.; Kornov, L., Driscoll, P.	Climate change in Danish planning system	Avoiding climate change uncertainties in Strategic Environmental Assessment	Danish	2013	Strategic thinking
Petrov, L.O.; Shahumyan, H.b; Williams, B.c.; Convery, S.d.	Application of scenario modeling and indicator evaluation for sustainable land use management in the Greater Dublin Region	Research article: Applying spatial indicators to support a sustainable urban future	Dublin	2013	Cellular automata (MODAL model) and ArcGis (FRAGSTAT)
Geneletti D.	Case study research aimed at empirically exploring how the implementation of different land-use zoning policies affect the future provision of a set of ecosystem services	Assessing the impact of alternative land-use zoning policies on future ecosystem services	Chile	2013	Trade-off scenario analysis
Morrissey J.;, Usha Iyer-Raniga; McLaughlin P.; Mills A.	A strategic overview of various project alternatives, taking account for stakeholder and expert input in infrastructure projects	A Strategic Project Appraisal framework for ecologically sustainable urban infrastructure	Australia	2012	Strategic Project Appraisal (SPA) framework
Liu, Y.; Guo, H.; Liu, H.; Yang, P.; Yang, Y.; Du, X.	An integrated framework for the evaluation of different projects for the area of Zhengzhou Airport	An integrated methodology framework for cumulative environmental assessment of regional development plan	China	2011	Scenario analysis and system dynamic in Cumulative environmental assessment

(continued)

Table 2 (continued)

Author	Decision context	Title	Case study	Year	Type of analysis
Bai, H.-T; Wang, H.-Z.; Xu, H.; Zhu, T.	The study develops an integrated assessment method based on accounting uncertainty of environmental impacts	Accounting for uncertainty in evaluating water quality impacts of urban development plan	China	2010	Integrated methodology for scenarios evaluation and development
Liu, Y.; Chen, J.; He, W.; Tong, Q.; Li, W.	Urban transformation in China	Application of an uncertainty analysis approach to strategic environmental assessment for urban planning	China	2010	Montecarlo analysis and HSY algorithm
Zhou, J.; Liu, Y.; Chen, J.; Yu.	Impact of urban transformation in water river	Uncertainty analysis on aquatic environmental impacts of urban land use change	China	2008	Scenario analysis, Stochastic simulation
Hojer, M.; Ahlroth, S.; Dreborg, K.-H., Ekvall, T. Finnveden, G.	Methodological paper on future studies in SEA	Scenarios in selected tools for environmental systems analysis	Sweden	2008	Predictive and normative Scenarios
Wang, J., Guo,H.; Liu, L.; Hao, M.; Zhang, M.; Lu, X.; Xing, K.	The presents the development of an inexact multi-objective programming (IMOP) model and its application to the strategic environmental assessment (SEA) for the regional development plan for the Hunnan New Zone (HNZ) in Shenyang City, China	Inexact multi-objective programming approach for strategic environmental assessment on regional development plan	China	2004	Multi-objective programming tool

The dis-homogeneity of the study examined in terms of theoretical approach, methods and results obtain suggests that there is a need for further research in this area including both methodology and practical case studies. On the base of the latter consideration in the next paragraph a cognitive methodological approach is presented with the aim to introduce a general framework for scenario analysis in SEA.

4 A Cognitive Methodological Framework

The development of scenarios is a complex process that includes different approaches and methodology and involves stakeholders and scientist's interaction and expert knowledge. On the same time SEA procedure is also a complex process that includes different phases: (a) screening and scoping; (b) development of an environmental report; (c) assessment of the environmental report and the outcome of the consultations; (d) the decision; (f) information on the decision; and (g) monitoring (article 11 of the Environmental Code). In each of these phases, scenario analysis can contribute to support the decision making process and reduce complexity both at spatial and temporal scale due to the strategic nature of SEA. Actually SEA is not only about technical studies but also about setting a platform for stakeholder dialogue and acting as a facilitator to decision problem. So the role of scenarios becomes more relevant to formulate, discuss and assess strategic option (Partidário 2012).

Here we propose a formal scenarios development (Fig. 2) to be used in SEA by

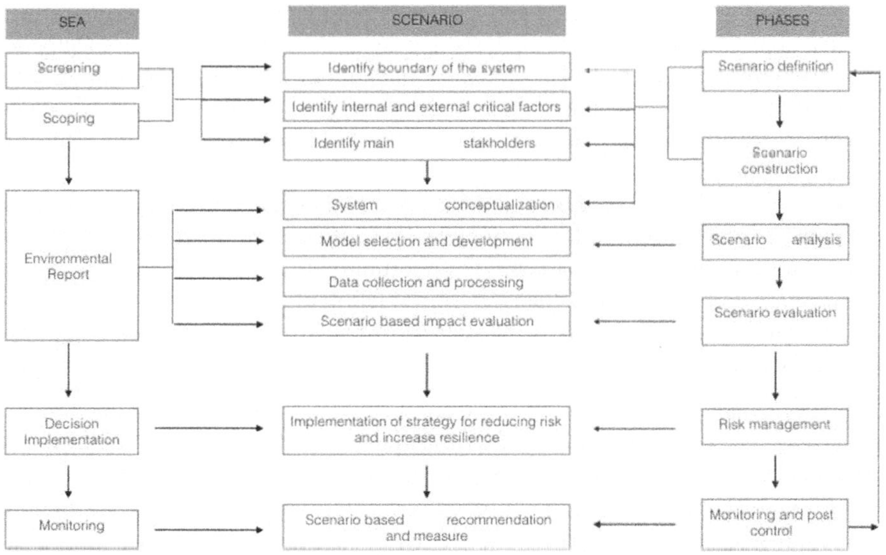

Fig. 2 Methodological framework

describing scenarios as an interaction process structured in six fundamental phases (Godet 2000) that are:

(1) *Scenario definition*: in this phase is important to set the basic line condition identifying the main boundary of the system, internal and external critical factors and key variable that can support the screening and scoping phase. The identification of the main the group of stakeholder to be involved is also a crucial activity. Actually effective scenario definition results from extensive discussion among stakeholders and researcher.
(2) *Scenario construction*: this phase comes after the scoping phase where an environmental report must be presented with all quantitative and qualitative information about future configuration. From a modeling based approach, scenario construction may consist in three major steps: system conceptualization that enhance and facilitate communication with stakeholders, capture key decision factor and verify the consistency of the hypothesis and the relationship on the base of qualitative and quantitative data and information.
(3) *Scenario analysis*: the focus is on identify consistency of the scenarios developed in the previous phase on the base of quantitative models to reduce uncertainty and underline notable system condition or behavior.
(4) *Scenario evaluation*: this phase is really important because the most plausible scenarios can be assessed on the base of different evaluation methods as Multicriteria analysis, Cost effective analysis, Life Cycle Assessment (LCA) etc.

    The role of evaluation methods appears very interesting in this field not only to support a ranking of choice scenarios, but also to guide decision makers in managing a decision process and its results through a descriptive or interpretative approach. Multicriteria analysis appear to be more appropriate to support the exploration of alternative scenarios, in fact they are capable of dealing with multiple dimension, soft data, interactive strategies, and try to give more attention to conflict arising among various stakeholders involved in the decision making process.
(5) *Risk management* is the phase of decision and implementation of strategies, for reducing vulnerability of risks, increasing resilience to problematic condition and positioning resources to exploit opportunity.
(6) *Monitoring phase* regards the ex post evaluation of the implemented strategies and review and correction of the scenarios developed checking the objectives reached in a circular mode.

As the Fig. 2 shown each phase of the scenario development approach can be integrated with the SEA procedure in order to support the entire decision-making in a sustainable and participatory manner. The decision process is extremely dynamic and move on a circular approach not based on a priori well defined choice structure between distinct possible alternatives, but can be interpreted as a process of building alternative scenarios and strategic actions where elements which determine the actions are in a moment or condition of cognitive equilibrium.

5 Conclusion

The paper presents a cognitive methodological framework based on scenario analysis to support SEA decision-making process in a participatory manner.

The proposed approach is based on scenario analysis as a useful tool to manage the complexity and the uncertainty of SEA. As argued by Godet scenario planning aim to identify key variable of a decision process, analyze actor game, reduce uncertainty on key questions and pick up the most favorable environmental scenarios using expert methods.

Although is strongly recognized in the literature the importance of strategic nature of SEA procedure in contrast to Environmental Impact assessment (EIA) approach at date still doesn't exist a common vision to cope with. As the literature review shows little case studies has been implemented where the term scenario is used in a consistent way. Scenario analysis is linked with the collective thinking within appropriate workshop useful to introduce future dimension and establish a complete diagnosis about the context with its environment, identify and assessing strategic choice option.

In this perspective there is a need for further research in this area including both methodology and practical case studies.

References

Amer M, Daim TU, Jatter A (2013) A review of scenario planning. Futures 46:23–40. Pergamon

Bai HT, Wang HZ, Xu H, Zhu T (2010) Accounting for uncertainty in evaluating water quality impacts of urban development plan. Environ Impact Assess Rev 30(4):219–228

Beck U (1999) World risk society. Polity Press, Cambridge UK

Bell W (1997) Foundations of futures studies, vol 1–2. Transaction publishers, New Bruns-wick, New Jersey

Bina O (2007) A critical review of the dominant lines of argumentation on the need for strategic environmental assessment. Environ Impact Assess Rev 27:585–606

Bragagnolo C, Geneletti D (2013) Dealing with land use decisions in uncertain contexts: a method to support Strategic Environmental Assessment of spatial plans. J Environ Plann Manage 57(1):50–77

Chermack TJ, Lynham SA (2002) Scenario planning: an examination of definitions and de-pendent variables. Hum Resour Dev Rev 1(3):366–383

Fowles J (1978) Handbook of futures research. Greenwood Press, Connecticut

Geneletti D (2013) Assessing the impact of alternative land-use zoning policies on future eco-system services. Environ Impact Assess Rev 40(1):25–35

Godet M (2000) The art of scenarios and strategic planning. Technol Forecast Soc Chang 65(1):3–22

Hojer M, Ahlroth S, Dreborg KH, Ekvall T, Finnveden G (2008) Scenarios in selected tools for environmental systems analysis. J Clean Prod 16(18):1958–1970

Inayatullah S (1996) Methods and epistemologies in futures studies. In: Slaughter R (ed) The knowledge base of futures studies, vol. 1. Futures Study Centre, Hawthorn Australia, pp 186–203

Jiliberto R (2011) Recognizing the institutional dimension of strategic environmental assess-ment. Impact Assess Proj Appraisal 29(2):133–140

Kahn H, Wiener A (1967) The year 2000: a framework for speculation on the next thirty. MacMillan, New York
Khakee A (1999) Participatory scenarios for sustainable development. Foresight 1(3):229–240
Kolb D (1984) Experiental learning. Prentice Hall, Englewood Cliffs, New Jersey
Lamers M, Liggett D., Tin T (2013) Strategic thinking for the Antarctic environment: the use of assessment tools in governance. In: Tin T, Liggett D, Maher P, Lamers M (eds) Antarctic futures: human engagement with the Antarctic environment. Springer
Larsen SV, Kornov L, Driscoll P (2013) Avoiding climate change uncertainties in Strategic Environmental Assessment. Environ Impact Assess Rev 43:144–150
Liu Y, Chen J, He W, Tong Q, Li W (2010) Application of an uncertainty analysis approach to strategic environmental assessment for urban planning. Environ Sci Technol 15;44(8):3136–3141
May GH (1996) The future is ours: foreseeing, managing and creating the future. Adaman-tine Press, London
Mietzner D, Reger G, (2005) Advantages and disadvantages of scenario approaches for strategic foresight. Int J Technol Intell Plann 1(2)
Morrissey J, Iyer-Raniga U, McLaughlin P, Mills A (2012) A strategic project appraisal framework for ecologically sustainable urban infrastructure. Environ Impact Assess Rev 33:55–65. Elsevier
Nijkamp P (1994) Global environmental change; management under long-range uncertainty. J Contingencies Crisis Manag 2(1):1–10
Noble BF, Gunn J (2015) Strategic environmental assessment. In: Hanna K (ed) Environmental assessment in Canada: practice and participation. Oxford University Press, Don Mills. ON
Noble B, Nwanekezie K (2016) Conceptualizing strategic environmental assessment: principles, approaches and research directions. Environ Impact Assess Rev. Elsevier
Partidario MR (2007) Strategic environmental assessment good practices guide: methodological guidance. Portuguese Environment Agency Amadora, Lisbon
Partidário MR (2012) Strategic environmental assessment better practice guide: methodological guidance for strategic thinking in SEA. Portuguese Environment Agency and Redas Energeticas Nacionais, Lisbon
Partidário MR (2015) A strategic advocacy role in SEA for sustainability. J Environ Assess Policy Manag 17(1):1–8 (1550015)
Petrov LO, Shahumyan H, Williams B, Convery S (2011) Applying spatial indicators to support a sustainable urban future. Urban Futures. 25, Cambridge
Ratcliffe J (2002) Scenario planning: strategic interviews and conversations. Foresight 4(1):19–30
Rozas-Vasquez D, Pena-Cortes F, Geneletti D, Rebolledo G (2014) Scenario modeling to support strategic environmental assessment: application to spatial planning of coastal wetlands in la Araucanía region, Chile. J Environ Assess Policy Manage 16(02)
Schwartz P (1991) The art of the long view. Planning for the Future in an Uncertain World, Doubleday Currency, New York
Sizo A, Noble B, Bell S (2015) Futures analysis of urban land use and wetland change in Saskatoon, Canada: an application in strategic environmental assessment. Sustainability 7(1):811–830
Slootweg R, Jones M (2011) Resilience thinking improves SEA: a discussion paper. Impact Assess Proj Appraisal 29(4):263–276
Torrieri F, Batà A (2017) Spatial multi-criteria decision support system and strategic environmental assessment: a case study. Buildings 7(4):96
Torrieri F, Nijkmap P (2008) Scenario analysis in spatial impact assessment: a methodological approach. In: Vreeker R, Deakin M, Curwell S (eds) Sustainable urban development. Volume 3: The toolkit for assessment. Routledge, Oxon
Van der Heijden K (1996) Scenarios: the art of strategic conservation. Wiley, Chichester, UK
Vleugel J (2000) Design of transport and land use scenarios—Principles and application. Ph.D. Thesis. Free University of Amsterdam, Amsterdam
von Reibnitz U (1988) Scenario techniques. McGraw Hill, New York
Wack P (1985) Scenarios: uncharted waters ahead. Harvard Bus Rev 63(5):72–89

Wang J, Guo H, Liu L, Hao M, Zhang M, Lu X, Xing K (2004) Inexact multi-objective programming approach for strategic environmental assessment on regional development plan. Prog Nat Sci 14(11):950–959
Wilson I (1978) Scenarios, handbook of futures research. In: Fowles J (ed) Greenwood Press. Connecticut, pp 225–248
Wood CM, Djeddour M (1992) Strategic environmental assessment: EA policies, plans and programmes. Impact Assess Bulletin 10:3–22
Zhou J, Liu Y, Chen J, Yu F (2008) Uncertainty analysis on aquatic environmental impacts of urban land use change. Front Environ Sci Eng China 2(4):494

Strategic Environmental Assessment (SEA) and Multi-Criteria Analysis: An Integrated Approach

Alessandra Oppio and Marta Dell'Ovo

Abstract With the aim of enhancing the level of sustainability of plans and programmes adopted by local, regional and national authorities, the European Commission (EU) has adopted the Directive 2001/42/EC on the assessment of effects of plans and programmes on the environment. Starting from the analysis of strengths and possible uses of Multi-Criteria Analysis (MCA) and the investigation of its application in combination with the SWOT Analysis and the Stakeholder Analysis, the paper aims at presenting a multi-methodological approach based on the use of MCA for Strategic Environmental Assessment (SEA). Given the spatial nature of the decision problem the multi-methodological approach is moreover combined with Geographic Information System (GIS). The Multicriteria-Spatial Decision Support System (MC-SDSS) proposed is able to support the decision-making processes in the field of environmental management by providing evidence and increasing the level of choices' transparency and legitimacy.

Keywords SEA · Multi-criteria analysis · Multi-methodological approaches

1 Introduction

The instance of improving choices' legitimacy and transparency in the field of environmental management, is one of the reason that has inspired the European Commission (EU) to adopt the Directive 2001/42/EC on the assessment of effects of plans and programmes on the environment. With the aim of enhancing the level of sustainability of plans and programmes adopted by local, regional and national authorities,

A. Oppio
Department of Architecture and Urban Studies DASTU, Politecnico Di Milano, Milan, Italy
e-mail: alessandra.oppio@polimi.it

M. Dell'Ovo (✉)
Department of Architecture, Built Environment and Construction Engineering ABC, Politecnico Di Milano, Milan, Italy
e-mail: marta.dellovo@polimi.it

G. Campeol (ed.), *Strategic Environmental Assessment and Urban Planning*, Green Energy and Technology, https://doi.org/10.1007/978-3-030-46180-5_4

the Strategic Environmental Assessment (SEA) Directive has defined a detailed procedure based on the following steps: (i) preparation of a report where possible effects on the environment are detected and solutions and strategies identified; (ii) public consultation about the proposal identified and the report; (iii) transboundary consultations and implementation of the proposal; (iv) decision phase where the results of the consultations are evaluated; (v) information about the decision taken and the proposal adopted; (vi) monitor of the effect on the environment; (vii) compliance of other legislations (Article 3).

Since a policy aimed at supporting citizens and stakeholder involved cannot be based only on producing evidence (De Marchi et al. 2016), but it should imply the use of evaluation methodologies able to guide in the decision-making process, it is evident as the SEA Directive is not the answer but it is a starting point. In fact, as already discussed by Partidario (2000) and Torrieri and Batà (2017), the SEA can be considered more as a steering than a prescriptive framework aimed at promoting good practices.

Given these premises a multi-methodological approach able to take into consideration all the aspects involved in the environmental assessment and to satisfy stakeholders with different and sometimes conflicting interests and visions is suggested. This kind of approach is aimed to provide a deep knowledge about the territory under investigation (Oppio et al. 2016; Dell'Ovo et al. 2018) by considering all the dimensions of the sustainability aimed at resulting with an overall evaluation of the project proposed. With this purpose, the study proposes an integration of Strategic Environmental Assessment (SEA) with Multi-Criteria Analysis (MCA) based on a comprehensive modelling both of the decision problem and of the decision context. In detail, the paper is divided into four sections. In the first one an overview about strengths of the MCA is given and possible integrations with other methodologies presented; the second is focused on investigating how other scholars have combined the use of MCA for SEA; the third part provides operational recommendations and the fourth draws the conclusions by putting in evidence the advantages given by the use of integrated decision support systems.

2 Multi-criteria Analysis and Decision Processes

Multi-criteria Analysis have been developed in order to support decision makers (DMs) and to help them to take better decisions (Roy 1990). The MCA has been considered as a revolution in the field of the Operational Research (Roy 1985) given by the possibility to study all the dimensions involved in the decision process and to evaluate them individually or as part of a unique system. The complexity given by the multidimensional nature of real-world problems, the need to involve stakeholders belonging to different categories and the instance of transparency and legitimacy of processes where the final decision is going to affect the whole community (Bonte et al. 1997, 1998; Janssen 2001), bring to light the importance of decision support systems able to guide the DMs. In particular, in the public sector and in the field of

SEA this task is even more important and many multi-dimensional models have been developed able to combine different disciplines and approaches. Below an overview of MCA's strengths is presented, underlining its possible uses with a focus on SEA.

2.1 Multi-criteria Analysis: An Overview

When does rise the necessity to be supported in taking a decision?

A decision problem exists when there is a difference between a current state and a desired state (Keeney 1996) and it is characterized by some key elements: (i) there are different actors involved with different opinions; (ii) more than one alternative is able to solve the issue defined; (iii) different criteria, both qualitative and quantitative, have to be taken into consideration to better describe and compare the potential courses of action under investigation.

Considering the first topic concerning the presence of (i) multi-stakeholders with different needs and expectations, according to de Almeida and Wachowicz (2017), these kind of decisions are more challenging compared to the individual ones, since in addition to the existence of conflicting objectives, different viewpoints and preferences have to be taken into account. The (ii) generations or identification of alternatives could be defined as another decision problem, since commonly, they are considered as "given" even if, in most of the cases, they are reviewed or defined during the process and not only at the beginning of it (Ozernoy 1985). Moreover, alternatives, in order to be evaluated, have to be well defined and described by the use of (iii) criteria. Criteria are the translation of relevant objectives, what it worth to be achieved by changing the current situation. The incomparability is detected when qualitative and quantitative criteria are both present in the decision framework or when the quantitative ones are described by different unite of measurement (u.m.) and then it becomes impossible to aggregate different performances (Janssen et al. 2000).

Given this premise, the MCA is considered as a strategic tool since it encompasses a series of techniques aimed at comparing alternative projects, by considering heterogeneous measures (Roy and Bouyssou 1995; Figueira et al. 2005) and evaluating at the same time different perspectives. Some of the most common steps of MCA are:

(a) the problem structuring that allows to identify an appropriate set of criteria by structuring and prioritizing the objectives detected (Roy 2005; de Almeida et al. 2016) by SWOT Analysis and Stakeholder Analysis;
(b) the generation of alternatives able to solve the problem previously modelled. Alternatives cannot be identified a priori but only after the definition of the objectives, otherwise shall be not satisfying (DCLG 2009; Keeney 1996);
(c) the standardization procedure that is able to transform incomparable criteria in a common, uniform and dimensionless scale, using (usually) a range from 0—the

worst performance—to 1—the best performance—in order to be easily evaluated. Many methods have been developed to perform this procedure (Hwang and Yoon 1981; Voogd 1983; Massam 1988; Malczewski and Rinner 2015) and it is important to underline how already in this phase the value functions resulted from the standardization represent the DM's preferences;

(d) the criteria weight elicitation aimed to take into consideration the different points of view of the stakeholders engaged in the decision process and to assign a different influence (weight) to criteria involved in the process according to their relevance in achieving the final aim of the evaluation (Riabacke et al. 2012);

(e) in relation to the decision context and to the typology of criteria considered, the aggregation allows to combine weights and standardized performance to result in overall values, for identifying the most suitable solution (choice problem), sorting (classification problem) or ranking (ordering problem) the available alternatives (Meyer and Roubens 2005; Malczewski and Rinner 2015). Many different aggregation rules exist and the main important difference is between compensatory and non-compensatory methods. In the first case, data are aggregated and negative performances are compensated by good performances. In the second case a threshold is defined for each criterion: if the performance does not satisfy it, the alternative under evaluation is then rejected.

2.2 How to Combine Multi-criteria Analysis Within Multi-methodological Approaches

Nowadays the Multi-criteria Analysis is more and more considered as an important procedure in combination with other methodologies to support DMs in structuring the decision problems (Marttunen et al. 2017) and in taking the final decisions. The SWOT—Strengths, Weaknesses, Opportunities, and Threats—Analysis, for example, is a strategic technique for planning since it gives the opportunity to recognize criticalities and potentials able to result in strategies. Grošelj et al. (2016) proposed a two-step approach based on SWOT analysis and Analytic Hierarchy Process (AHP—Saaty 1980) for the forest management and in detail for the comparison of possible future scenarios. Also Miyamoto et al. (2014), for the flood risk management, has combined MCA and AHP-SWOT to rank interventions and evaluate the most urgent one by developing quantitative methodologies.

Since different stakeholders are involved in the process and are affected by it, it is important to identify since the early stages of the decision process which actors are going to participate and which is their role. Della Spina (2018), with the aim of designing complex urban scenarios in support of strategic planning and urban regeneration, has integrated a multi-dimensional and multi-level approach. In particular, to define the most suitable alternatives has combined the MCA with the Stakeholder Analysis in order to elicit objectives and values of the actors involved. Ianni and

Geneletti (2010) to select forest restoration priority areas has integrated the Stakeholder Analysis with the MCA too. The Stakeholder Analysis brought two benefits in the case study analysed by the paper, the first one concerns the identification of all actors involved in forest use and management and the second one is the description of expectations of beneficiaries of Forest Landscape Restoration.

Another possible implementation of the MCA regards the combination with Geographic Information System (GIS). For example, Singh et al. (2018) integrated these two methodologies to delineate groundwater potential and, in detail, GIS allowed to handle a large amount of spatial data. Moreover, Torrieri and Batà (2017) have proposed to combine GIS and MCA to contribute to the SEA from a methodological perspective and to support the generation of urban planning scenarios. The further support detected by the use of spatial data concerns the evaluation of impacts under a multi-dimensional point of view and the possibility to map the results of the analysis with a comprehensive and clear representation even for non-expert actors.

3 The Use of the Multi-criteria Analysis for Strategic Environmental Assessment (SEA)

According to what it emerges from the previous section, the use of the MCA in the field of the SEA could be strategic in order to provide a deep knowledge of the decision context, to elicit objectives and expectations of stakeholders involved in the process—both direct and indirect—and to evaluate potential plans and programms.

At this stage a literature has been framed and, as suggested by Prasara and Gheewala (2017) and Moghadam et al. (2017), it has been structured according to the following four-stages:

1. "*Literature search*": selection of a database to develop the analysis;
2. "*Screening process*": selection of keywords to narrow the analysis;
3. "*Selection of literature*": selection of papers according the aim of the analysis;
4. "*Including literature*": selection of data to detect by the analysis of the papers.

Considering the framework proposed, (1) the Scopus database has been used and the research has been based on the selection of the following (2) keywords "Strategic Environmental Assessment" and "Multi-criteria Analysis" or "MCA" to narrow the analysis. (3) 18 documents have been identified and judged suitable, according to the title and abstract, to be further studied. The analysis has been focused on (4) understanding in which SEA context the MCA has been developed; which decision problems have been faced and if within the decision process have been combined the methodologies described in the previous section: SWOT; Stakeholder Analysis and GIS (Table 1).

Bobylev (2006) has evaluated the environmental impacts of Urban Underground Infrastructure (UUI) development policies using the MCA and stressing the necessity to incorporate it in cities' masterplans. Also other scholars have investigated the same

Table 1 Literature review

Author	Year	Country	Context	Decision problem	SWOT	Stakeholder analysis	GIS
Bobylev N.	2006	Japan	Infrastructure	Multi-criteria analysis (MCA) for evaluating environmental impacts of UUI development polices	No	Yes	No
Celik M. et al.	2008	Turkey	Environmental Management	Approach based on Fuzzy Axiomatic Design (FAD) to assign suitable MCA technique during execution process of ANSEA.	No	No	No
Olazabal M. et al.	2010	Spain	Environmental Management	Integrated approach to sustainable spatial management based on the development of an integrated decision-support system	No	Yes	Yes
Cerreta M. and Toro P.D.	2012	Italy	Planning policy	The ISA approach to improve the transparency of evaluation process and evaluate planning choice for Cava de' Tirreni (Italy)	Yes	Yes	Yes
Lanado E. et al.	2010	Italy	Participatory process	Presentation of AMACI software to perform the Strategic Environmental Assessment of the local planning process of the Municipality of Trezzo sull'Adda (Italy)	No	Yes	No

(continued)

Table 1 (continued)

Author	Year	Country	Context	Decision problem	SWOT	Stakeholder analysis	GIS
Garfì M. et al.	2011	Brazil	Environmental Management	Contribution of AHP in two stages of SEA procedure applied to water programmes in developing countries	Yes	Yes	No
Azzellino A. et al.	2011	Italy	Renewable Energy	MCA to evaluate the best location for Wave Energy Converter installations	No	No	Yes
Schetke S. et al.	2012	Germany	Environmental Management	Innovative MCA of greenfield and infill sites to evaluate their sustainability and resource efficiency	No	Yes	Yes
White L and Noble B.	2012	Canada	Energy	A strategic environmental assessment (SEA) framework for electricity sector planning is developed and applied to evaluate electricity supply scenarios	No	Yes	No

(continued)

Table 1 (continued)

Author	Year	Country	Context	Decision problem	SWOT	Stakeholder analysis	GIS
Buchala Bicca Oliveira et al.	2012	Brazil	Sectoral Planning	Determining an Index of Sustainability of Expansion of the Sugar and Alcohol Sector (IScana), employing the fuzzy logic and AHP to aggregate the indicators generated in the baseline step of the SEA process, in order to consolidate a sector monitoring tool.	No	No	No
Vukicevic J.S. and Nedovic-Budic Z.	2012	Ireland	Planning policy	Integration of SEA into the planning process based on GIS multi-criteria analysis to identify potential impacts of planned activities on the environment	Yes	No	Yes
Naddeo V. et al.	2013		Environmental Management	Innovative integrated methodology for SEA, able to implement engineering aspects – mostly related to forecasting models and evaluation of impacts on the environment – as well as social issues, as to understand economic implications	Yes	Yes	No

(continued)

Table 1 (continued)

Author	Year	Country	Context	Decision problem	SWOT	Stakeholder analysis	GIS
Thompson UC et al.	2013	Canada	Infrastructure	Investigating the level of harmonization between SEA, PPP and proposed projects and their possible alternatives (CAPPP) analysing the Turcot Interchange redevelopment project	Yes	Yes	No
Capolongo S. et al.	2016	Italy	Health	Approaches to include health issues into land use plans and urban development projects and development of a tool able to provide hygiene and health evaluation of urban plans	No	No	No
Sizo A. et al.	2016	Canada	Planning policy	An approach to SEA to support PPP development and implementation for urban wetland conservation	No	No	No
Karlson M. et al.	2016	Sweden	Infrastructure	Methods for the integration of important ecological and geological sustainability criteria for planning of transport infrastructure corridors in a systematic and transparent way	No	Yes	Yes

(continued)

Table 1 (continued)

Author	Year	Country	Context	Decision problem	SWOT	Stakeholder analysis	GIS
Sharma S. and Geerlings H.	2017		Infrastructure	New methodological approach Sustainability Benefits Assessment in Urban Transport Project Appraisal (SBA-UT),	No	No	Yes
Torrieri F. and Batà A.	2017	Italy	Environmental Management	The potentiality of the ISMDSS to evaluate the impacts of different scenarios with the aim of developing a sustainable urban municipal plan	No	Yes	Yes

context. Thompson et al. (2013) tried to solve the gap between SEA, governmental plans, policies and programmes (PPP), and environmental impact assessments (EIAs) by proposing a Compliance Analysis for PPP (CAPPP) method to be applied to the Turcot Interchange redevelopment project. Karlson et al (2016) used the spatial multi-criteria analysis (SMCA) techniques for planning railway corridors based on ecological and geological criteria, while Sharma and Geerlings developed a new approach based on a systematic assessment of sustainability benefits of a project. Moving to the environmental management field, Celik et al. (2008) developed the Analytical Strategic Environmental Assessment (ANSEA) framework to overcome the criticalities detected in the traditional SEA procedure, with a special attention to the strategic decision-making levels. In fact, the ANSEA operational framework provides a contribution to the formulation of policies, programs and planning activities and it has been integrated with Fuzzy Information Axiom (FAD) methodology to select suitable MCA techniques in relation to the problem. Even if the decision context is the same, Garfì et al. (2011) has used MCA (in detail the AHP) for strategic environmental assessment of water programmes in Brazil, while Schetke et al. (2012) to develop sustainable strategies of housing development. Naddeo et al. (2013) has focused the attention on framing an integrated approach for SEA. Still changing the decision context, Olazabal et al. (2010) and Laniado et al. (2010) stressed more the participatory processes beyond the sustainable decision-making process management. Their tools, in fact, are aimed to involve from the early stage different levels and categories of stakeholders improving the communication and the transparency of the overall evaluation process. In the first case the purpose was to evaluate the urban plan of a sector under development located in the province of Araba (Spain), while in the second case the purpose was to support the SEA of the Town-Planning Scheme of the Municipality of Trezzo sull'Adda (Italy). Finally, another interesting aspect to be taken into consideration in the definition of SEA is the health issue, in particular for urban development plans. In fact, for Capolongo et al. (2016), urban planning is a form of risk prevention and nowadays SEA procedures rarely consider this perspective.

From the literature it is possible to underline that:

- many fields are covered by the SEA procedure;
- the MCA is considered by most of the selected studies as an important support to evaluate possible development scenarios and in particular their effects;
- this kind of procedure is open to engage citizens and stakeholders and, thus, to improve the transparency of decision-making processes;
- most of the analysed papers involves the use of GIS;
- only few scholars previously have combined the SWOT Analysis within the SEA procedure;
- the Stakeholder Analysis is considered as a fundamental phase to deeply understand the decision problem since allows to clarify the role played by different actors and their expectations.

4 Operational Recommendations to Develop a Multi-methodological Approach for SEA

Once analyzed the literature review and understood the advantages of combining the MCA with other methodologies, it is possible to develop a Multi-methodological Approach and operational recommendations to frame the SEA procedure. In detail, a flowchart is proposed in order to better explain the different steps aimed to strengthen its potentials as an integrated decision support system (Fig. 1).

From Fig. 1 it is possible to deduce how a great importance has been assigned to the preparatory activities leading to the evaluation as the SWOT Analysis and the Stakeholder Analysis. Both of them contribute to develop a cognitive framework and to elicit fundamental objectives in order to solve the decision problem.

The complexity given by the presence of multiple and sometimes conflicting objectives in urban transformation, and the necessity to take into consideration at the same time needs of actors directly or indirectly involved in the decision problem, suggest to frame a Multicriteria-Spatial Decision Support System (MC-SDSS) able to combine the potential of GIS—collecting, elaborating and representing on the map spatial data—with those of MCA—able to support decision-making processes through the elicitation of both qualitative and quantitative objectives and to evaluate possible impacts of the decisions taken (Malczewski 1999). MC-SDSSs allow to consider at the same time different territorial dimensions—economic, environmental, social, etc.—and to visualize them at a spatial level, in order to structure and manage the decision problems concerning integrated planning. A system based on criteria, sub-criteria and indicators spatially represented by a GIS software, opportunely standardized and aggregated according to the importance in achieving the final aim, allows to result with a synthesis map able to show potentials and the critical aspects of the territorial context under investigation. The outputs (Suitability maps) are directly connected with inputs selected according to the emerging complexity of the territory.

Suitability maps, once standardized on the basis of adequate value functions, for example, are able to point out which area is the more appropriate for the location of

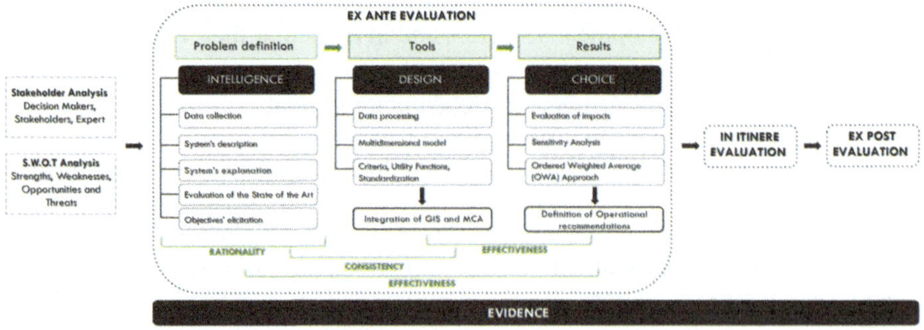

Fig. 1 Multi-methodological flowchart

services, facilities or infrastructures. It is important to underline how the strength of suitability maps is given by the possibility to read the total result (aggregated) and the partial ones (disaggregated) according to the structure of the decision problem. By providing different layers of knowledge, it is easier to comprehend precisely where are weaknesses and how to mitigate them.

Given these premises and with a special attention to strategic planning, the integration of GIS and MCA represents an effective support in the SEA field.

Considering the Multi-methodological Approach proposed, it is composed by the following phases: 1. Intelligence; 2. Design; 3. Choice. How it is possible to appreciate from Fig. 1, there is a flow of activities through the three stages and all the phases of the decision-making process involve the methodological contribution of both GIS systems and MCA techniques, in addition to SWOT Analysis and Stakeholder Analysis.

In detail, the phase 1. Intelligence represents the structure of the problem, being the system described and objectives elicited. In this context it is possible to identify criteria able to describe and to achieve the objectives detected according to the needs and expectation of stakeholders (Stakeholder Analysis) and criticalities of the territory previously analysed (SWOT Analysis) (Keeney 1992). Still within this preliminary phase of analysis, from the data collection and the system's description, problems to be solved or opportunities to take advantage of emerge (Sharifi and Rodriguez 2002). Data collected could be vector or raster and they are visualized through specific "Source Maps".

The phase of 2. Design is based on the data processing and it is aimed to develop the multi-criteria structure of the problem through the definition of the relation between objectives, attributes and DM's preferences (Malczewski 1999). The result is the "Criterion Maps". An important role is played by the standardization procedure—to make information comparable—and the criteria weight elicitation—to assign a different influence according to the purpose of the analysis. Given the spatial nature of the decisional context, at this stage, it is provided the integration of MCA with GIS systems and the definition or generation of possible alternatives and/or solutions is considered.

During the last phase—3. Choice—"Criterion Maps" are aggregated to result in "Suitability Maps" and then alternatives are subjected to evaluation. It is very useful to develop a sensitivity analysis in order to test the robustness of the model and to obtain operational recommendations. Given the multidimensional nature of the decision problem, it is strategic to evaluate the impact of each alternative defined, according to the dimensions analysed, to have a comprehensive evaluation of each scenario and, at the end, to take a decision consistently with the objectives elicited during the first phase.

Furthermore, the application of the Ordered Weighted Average (OWA) (Yager 1988) method allows to validate the feasibility of the strategies proposed, since the trade-off among criteria is considered and thresholds of acceptability of the risk are defined. The trade-off consists in evaluating to what extent a criterion can compensate another, while the risk may be defined as the probability that the decision taken is wrong. The OWA approach used in this phase becomes strategic to generate and visualize unexpected solutions and predictive scenarios (Ferretti and Pomarico 2013).

"Evidence" crosses all the phases of the multi-methodological approach and is related to the total amount of data and information available for the DM. The Evidence can be based on facts, values, knowledge or experiences, and represents a key resource in all the phases of the decision-making process.

5 Conclusions

Considering the framework proposed and the discussion provided about the concept of Evidence, Sanderson (2002) argues about the presence of two different forms of evidence. The first one considers the effectiveness, the achievement of practical results by the work of the government, while the second is more focused about the relevance of policies and how they can work in different contexts based on specific levels of knowledge and aimed to improve social systems.

At the same time two different approaches exist to pursue the evidence (Fischer 2003; Stanhope and Dunn 2011). The first strategy concerns the proposal of transparent methodologies supported by the use of tools able to strengths phases of the decisions, the second one considers the involvement of citizens and their participation in decision-making processes by the elicitations of their expectations and interests.

By the adoption of one of the approaches proposed the final goal is the achievement of the satisfaction of actors involved and affected by the policy proposed. Also the satisfaction can be defined in two ways, both as a result and as process (Yi 1990; Grigoroudis and Siskos 2009). In the first case the satisfaction is obtained at the end of the process, as an outcome, while in the second is given by the process in relation to the perceptive feeling of fulfilment.

The concepts discussed and described in this paper suggest the definition of the evidence-based decision making (EBDM), able to synthesize the idea of evidence, defined as a fair relationship between stakeholder and DM (to brings awareness about the decision process), and as the possibility to create policies based on knowledge (De Marchi et al. 2016). The difficulty in design evidence-based policies is mainly due to the long time horizon to be effective and then measured (Sanderson 2002; De Marchi et al. 2016), even more exacerbated by delays of bureaucratic procedures.

The framework proposed in the previous section, aims to properly pursue the objectives of the EBDM by applying both strategies direct to achieve the evidence (the use of transparent tools and the participation of citizens to the decision). In fact, the whole process is supported by tools and consolidated methodologies and it is based on a deep understanding and analysis of the key actors. Within the multi-methodological approach, moreover, the concept of Evidence and satisfaction is conceived as the "result" obtained at the end of the application and, at the same time, as a "process" obtained during the development of these three phases described—Intelligence, Design and Choice—given by their complete comprehension. These kind of multi-methodological approaches, to be adopted, should be promoted by the internal bodies of the government and integrated in planning and management policies in terms of "good practices" (Sanderson 2002).

References

Azzellino A, Lanfredi C, Contestabile P, Ferrante V, Vicinanza D (2011) Strategic environmental assessment to evaluate WEC projects in the perspective of the environmental cost-benefit analysis. In: The twenty-first international offshore and polar engineering conference. International Society of Offshore and Polar Engineers

Janssen R, Van Herwijnen M, Beinat, E (2000) DEFINITE for Windows. A system to support decisions on a finite set of alternatives (Software and package and user manual)

Bobylev N (2006) Strategic environmental assessment of urban underground infrastructure development policies. Tunn Undergr Space Technol 21(3–4)

Bonte RJ, Jvd Burg, Janssen R, Mooren RHJ, deSmidt JT (1997) Notitie o6er multi-criteria analyse inmilieu-effectrapportage. Commissie voor de milieu-effectrapportage, Utrecht

Bonte RJ, Janssen R, Mooren RHJ, Smidt JTd, Burg JJvd (1998) Multicriteria analysis: making subjectivity explicit. In: New experiences on environmental impact assessment in the Netherlands: process, methodology, case studies, Commissie voor de milieueffectrapportage van M (ed) Utrecht, Commissie voor de milieueffec-trapportage, pp 23–28

Buchala Bicca Oliveira D, Rodrigues JP, Ferreira da Silva L, Tarso Sanches Oliveira P (2012) Multi-criteria analysis in the strategic environmental assessment of the sugar and alcohol sector. Acta Scientiarum Technol 34(3)

Capolongo S, Lemaire N, Oppio A, Buffoli M, Roue Le Gall A (2016) Action planning for healthy cities: the role of multi-criteria analysis, developed in Italy and France, for assessing health performances in land-use plans and urban development projects. Epidemiol Prev 40(3–4):257–264

Celik M, Cebi S, Ucal I (2008) Enhancement of multi-criteria analysis within analytical strategic environmental assessment. In: 2008 IEEE International engineering management conference (IEMC Europe 2008). IEEE, pp 1–4

Cerreta M, Toro PD (2010) Integrated spatial assessment for a creative decision-making process: a combined methodological approach to strategic environmental assessment. Int J Sustain Dev 13(1–2):17–30

de Almeida AT, Almeida JA, Costa APCS, Almeida-Filho AT (2016) A new method for elicitation of criteria weights in additive models: flexible and interactive tradeoff. Eur J Oper Res 250(1):179–191

de Almeida AT, Wachowicz T (2017) Preference analysis and decision support in negotiations and group decisions. Group Decis Negot 26:649–652. https://doi.org/10.1007/s10726-017-9538-6

De Marchi G, Lucertini G, Tsoukiàs A (2016) From evidence-based policy making to policy analytics. Ann Oper Res 236(1):15–38

Dell'Ovo M, Capolongo S, Oppio A (2018) Combining spatial analysis with MCDA for the siting of healthcare facilities. Land Use Policy 76:634–644

Della Spina L (2018) A multi-level integrated approach to designing complex urban scenarios in support of strategic planning and urban regeneration. In: International symposium on new metropolitan perspectives. Springer, Cham, pp 226–237

Ferretti V, Pomarico S (2013) Ecological land suitability analysis through spatial indicators: an application of the analytic network process technique and ordered weighted average approach. Ecol Ind 34:507–519

Figueira J, Greco S, Ehrgott M (2005) Multiple criteria decision analysis: state of the art surveys. Springer, New York

Fischer F (2003) Reframing public policy: discursive politics and deliberative practices. Oxford University Press
Garfì M, Ferrer-Martí L, Bonoli A, Tondelli S (2011) Multi-criteria analysis for improving strategic environmental assessment of water programmes. A case study in semi-arid region of Brazil. J Environ Mange 92(3):665–675
Grigoroudis E, Siskos Y (2009) Customer satisfaction evaluation: methods for measuring and implementing service quality, vol 139. Springer Science & Business Media
Grošelj P, Hodges DG, Stirn LZ (2016) Participatory and multi-criteria analysis for forest (ecosystem) management: a case study of Pohorje, Slovenia. For Policy Econ 71:80–86
Hwang CL, Yoon K (1981) Multiple attribute decision making: methods and applications. Springer, Berlin
Ianni E, Geneletti D (2010) Applying the ecosystem approach to select priority areas for forest landscape restoration in the Yungas, Northwestern Argentina. Environ Manage 46(5):748–760
Janssen R (2001) On the use of multi-criteria analysis in environmental impact assessment in The Netherlands. J Multi-criteria Decis Anal 10(2):101–109
Karlson M, Karlsson CSJ, Mörtberg U, Olofsson B, Balfors B (2016) Design and evaluation of railway corridors based on spatial ecological and geological criteria. Transp Res Part D: Transp Environ 46:207–228
Keeney RL (1992) Value-focused thinking: a path to creative decision analysis
Keeney RL (1996) Value-focused thinking: identifying decision opportunities and creating alternatives. Eur J Oper Res 92(3):537–549
Laniado E, Luè A, Muratori S (2010) Participation in multi-criteria decisions: a software tool and a case study. In: International congress on environmental modelling and software, p 321
Malczewski J (1999) GIS and multicriteria decision analysis. Wiley
Malczewski J, Rinner C (2015) Multicriteria decision analysis in geographic information science. Springer, New York, pp 220–228
Marttunen M, Lienert J, Belton V (2017) Structuring problems for multi-criteria decision analysis in practice: a literature review of method combinations. Eur J Oper Res 263(1):1–17
Massam BH (1988) Multi-criteria decision making (MCDM) techniques in planning. Prog Plann 30(1):1–84
Meyer P, Roubens M (2005) Choice, ranking and sorting in fuzzy multiple criteria decision aid. In: Multiple criteria decision analysis: state of the art surveys. Springer, New York, NY, pp 471–503
Miyamoto M, Osti R, Okazumi T (2014) Development of an integrated decision-making method for effective flood early warning system. J Disaster Res 9(1):55–68
Moghadam ST, Delmastro C, Corgnati SP, Lombardi P (2017) Urban energy planning procedure for sustainable development in the built environment: a review of available spatial approaches. J Clean Prod 165:811–827
Naddeo V, Belgiorno V, Zarra T, Scannapieco D (2013) Dynamic and embedded evaluation procedure for strategic environmental assessment. Land Use Policy 31:605–612
Olazabal M, Urzelai A, García G, Herranz-Pascual K, Abajo B, Feliú E, Santa-Coloma O, Aspuru I (2010) Sustainable spatial management: an integrated approach. In: Proceedings of the institution of civil engineers-municipal engineer, vol 163, no 1. Thomas Telford Ltd., pp 33–41
Oppio A, Buffoli M, Dell'Ovo M, Capolongo S (2016) Addressing decisions about new hospitals' siting: a multidimensional evaluation approach. Annali dell'Istituto superiore di sanita 52(1):78–87
Ozernoy VM (1985) Generating alternatives in multiple criteria decision making problems: a survey. In: Decision making with multiple objectives. Springer, Berlin, Heidelberg, pp 322–330
Partidario MR (2000) Elements of an SEA framework—improving the added-value of SEA. Environ Impact Assess Rev 20(6):647–663
Prasara AJ, Gheewala SH (2017) Sustainable utilization of rice husk ash from power plants: a review. J Clean Prod 167:1020–1028
Riabacke M, Danielson M, Ekenberg L (2012) State-of-the-art prescriptive criteria weight elicitation. Adv devis Sci

Roy B (1985) Méthodologie multicritère d" aide à la décision. Economica
Roy B (1990) The outranking approach and the foundations of ELECTRE methods. In: Readings in multiple criteria decision aid. Springer, Berlin, Heidelberg, pp 155–183
Roy B, Bouyssou D (1995) Aide multicritére à la décision: méthodes et case. Economica, Paris
Roy B (2005) Paradigms and challenges. In Multiple criteria decision analysis: state of the art surveys. Springer, New York, NY, 3–24
Saaty TL (1980) The Analytic Hierarchy Process. McGraw-Hill, New York
Sanderson I (2002) Evaluation, policy learning and evidence-based policy making. Public Adm 80(1):1–22
Schetke S, Haase D, Kötter T (2012) Towards sustainable settlement growth: a new multi-criteria assessment for implementing environmental targets into strategic urban planning. Environ Impact Assess Rev 32(1):195–210
Sharifi MA, Rodriguez E (2002) Design and development of a planning support system for policy formulation in water resource rehabilitation. J Hydroinformatics 4(3):57–175
Sharma S, Geerlings H (2017) Sustainability benefits assessment in urban transport project appraisal: a new method of transport project appraisal. In: International conference on smart and sustainable planning for cities and regions. Springer, Cham, pp 185–203
Singh LK, Jha MK, Chowdary VM (2018) Assessing the accuracy of GIS-based multi-criteria decision analysis approaches for mapping groundwater potential. Ecol Ind 91:24–37
Sizo A, Noble B, Bell S (2016) Connecting the strategic to the tactical in SEA design: an approach to wetland conservation policy development and implementation in an urban context. Impact Assess Proj Appraisal 34(1):44–54
Stanhope V, Dunn K (2011) The curious case of housing first: the limits of evidence based policy. Int J Law Psychiatry 34(4):275–282
Thompson UC, Marsan JF, Fournier-Peyresblanques B, Forgues C, Ogaa A, Jaeger JA (2013) Using compliance analysis for PPP to bridge the gap between SEA and EIA: lessons from the Turcot interchange reconstruction in Montréal, Québec. Environ Impact Assess Rev 42:74–86
Torrieri F, Batà A (2017) Spatial multi-criteria decision support system and strategic environmental assessment: a case study. Buildings 7(4):96
Voogd H (1983) Multicriteria evaluation for urban and regional planning. Pion Limited, London
Vukicevic JS, Nedovic-Budic Z (2012) GIS based multicriteria analysis in integration of SEA process into planning, case study: south west region, Republic of Ireland. Int J Environ Res 6(4):1053–1066
White L, Noble B (2012) Strategic environmental assessment in the electricity sector: an application to electricity supply planning, Saskatchewan, Canada. Impact Assess Proj Appraisal 30(4):284–295
Yager RR (1988) On ordered weighted averaging aggregation operators in multicriteria decision-making. IEEE Trans Syst Man Cybern 18(1):183–190
Yi Y (1990) A critical review of consumer satisfaction. Rev Mark 4(1):68–123

An Application of the Analytic Hierarchy Process (AHP) in the SEA Process of a Cross-Border Transport Strategy: The Veneto-Austria Corridor

Giovanni Campeol, Sandra Carollo, and Nicola Masotto

Abstract The meaning of Strategic Environmental Assessment is investigated by highlighting the methodological features of the model and of the different application stages. In this framework, the attention is focused on the Analytic Hierarchy Process—AHP and on its integration in the SEA process. Then the AHP application case is shown in the ex-ante stage of the SEA, that is the evaluation of a cross-border transport strategy in the Italy-Austria corridor.

Keywords SEA Analytic Hierarchy Process (AHP) · Cross-border transport strategies

1 The Meaning of the Strategic Environmental Strategy

The need to evaluate the effects on the environment—as a container of abiotic, biotic and human components—, arising from the implementation of actions defined by a planning or programming tool, had been consciously existing well before European Directive 2001/42/EEC was issued. Indeed, this Directive, especially since its implementation in the Member States legislation, focused the interest of the academic and technical-institutional communities on the evaluation issues and it was—and still is—subject to theoretical-methodological reflections and experimental applications at different scales, through the codification of the Strategic Environmental Assessment (SEA) process and the definition of environmental assessment methods.

Currently, the SEA represents the most articulated and exhaustive systemic assessment process to evaluate the consequences of the actions established by the

G. Campeol (✉)
Università Iuav di Venezia, Venice, Italy
e-mail: giovanni.campeol@gmail.com

S. Carollo · N. Masotto
Università degli Studi di Padova, Padua, Italy
e-mail: dott.sandracarollo@studiocaldato.it

N. Masotto
e-mail: n.masotto@gmail.com

G. Campeol (ed.), *Strategic Environmental Assessment and Urban Planning*, Green Energy and Technology, https://doi.org/10.1007/978-3-030-46180-5_5

Plans/Programs on the environment, and defines the environmental sustainability of the territorial transformation processes. In this sense, the application of the SEA, as a helpful tool to support decisions, influences the approach to territorial planning, and modifies its perspective. The need for a continuous methodological, technical and participatory dialogue between the SEA and the Plan engenders virtuous feedbacks in the elaboration process and its different implementation stages.

In fact, differently from the older project evaluation (EIA), the SEA does not fall within an authorization process; rather, it is a structural and endogenous part of the planning process and acquires the character of internal self-evaluation (inside the procedures) by using different assessment methods in relation to the specific Plan stages. Therefore, it is possible to make a parallelism between the life cycle of the Plan action—gestation, implementation and conclusion—and the temporal representation of the state of the environment in various moments realized by the SEA, which is articulated in three stages (Campeol and Carollo 2003):

- Ex ante stage, before the definition of the Plan/Program, in order to define the criteria on the basis of the criticalities and potentials emerging from the analysis of the environmental system, with the aim to direct alternative transformation assumptions and relevant guidelines;
- Ongoing stage, in parallel with the elaboration of the Plan/Program, by making an estimation of the environmental impacts produced by the Plan choices and their coherence with the orientations coming from the environmental assessment;
- Ex post stage, after the implementation of the Plan/Program, by verifying the soundness of the estimations made and, consequently, through monitoring, the real sustainability of territorial transformations.

The ex-ante stage is particularly interesting in terms of evaluation since it provides clear and precocious orientations to the transformation processes and helps the construction of Plan alternatives (scenarios).

2 The Contribution of the Analytic Hierarchy Process (AHP) to the SEA

An evaluation methodology that provides an important contribution to decision-making among various alternatives in front of multiple criteria is the Analytic Hierarchy Process—AHP (Saaty 1980, 1986), which defines the action priorities and intervention strategies, and is therefore an efficient and helpful tool in the first stage of the definition of the Plan strategies.

In terms of the classification of the Analytic Hierarchy Process within the different evaluation families, it can be fully included in the Multi-Criteria Analysis (MCA) methodologies suitable for evaluating the alternatives characterized by multiple and conflictual criteria—similarly to the SEA—, and especially to the sub-category of the

Multiple-criteria Decision-Making (MCDM) and Multi-criteria Decision Analysis (MCDA).

The Analytic Hierarchy Process (AHP) is one of the most commonly used methodologies within the MCDA for solving decision problems. It is a process of hierarchical measurement fit for defining the weights and measures of a set of options (also of planning type) by the stakeholders that participate in the evaluation process through a Working Table—WT (Masotto 2018). The process is based on the importance given to criteria and sub-criteria established in relation to a given strategic objective. Through the application of the AHP, it is possible to investigate alternative scenarios of large-scale territorial development (Campeol et al. 2016), enabling the evaluation of priority actions and intervention strategies (Saaty 1980, 1986; Campeol et al. 2016). Today, there exist many examples of the method applications to the evaluation problems in very diverse areas (Golden et al. 1989; Giangrande 2008; Caramia 2009).

These criteria are measured through a pairwise comparison, and their degree of importance is defined by the preference expressed by using the "semantic scale of Saaty". One of the peculiar elements of the Decision Making Theory—i.e. the study through mathematical and statistical methods of the decision to be preferred among varied possible alternatives—, is in fact the definition of weights for a set of different actions; coherently the AHP, through the identification of criteria and sub-criteria and their pairwise comparison, permits to give priorities to a set of decisional alternatives, comparing qualitative and quantitative evaluations, otherwise not directly comparable, and combining multidimensional measure scales in a single priority scale (Masotto 2018).

The AHP makes possible evaluating the preferability of alternative territorial development scenarios compared with defined criteria and in relation to the goal identified. The evaluation process is structured in consequential stages and permits to attribute weighting values to the different environmental components, through the use of a spreadsheet specifically created with a reliable open-source computing software able to generate clear responses (Campeol et al. 2017).

3 Application of the AHP: The Evaluation of a Cross-Border Transport Strategy Italy-Austria

The case subject to the proposed AHP application concerns different hypotheses of infrastructural development scenarios for the direct alpine connection between two weak areas, the first in Italy—the Veneto Region with the Province of Belluno—the second in Austria—Tyrol, with the District of Lienz. In Italy, the Veneto Region is the only alpine region without a road and/or railway pass to the North. This absence, combined with a general situation of demographic de-growth and economic impov erishment of the whole province, has triggered the need to assume infrastructural development hypotheses defined by alternative scenarios, set within the more general Trans-European Transport Network (TEN-T) and the territorial relationships of the Italian alpine area and the adjacent Dolomites territories (Fig. 1). This has enabled confirming the analyses carried out for the elaboration of the Strategic Plan and the related SEA of the Belluno Province (2007), which highlighted the main criticalities of the Belluno Province, especially of its infrastructures. In fact, the analyses performed in the large geographical area of the Province of Belluno irrefutably show how mobility and its necessary infrastructures need being implemented and intelligently "networked". Moreover, the definition of alternative scenarios of infrastructural development needs the application of the evaluation methodology able to prioritize objectives and criteria, with the aim to define a preferability scale of the different transformation scenarios identified (in this case, all characterized by the presence of a direct intra-alpine connection).

In defining the evaluation method, it is necessary to establish the importance of a given action compared with another, in relation to the decision criteria previously defined and shared (in whole or in part) by the public and private actors involved. The AHP needs prioritizing the objectives in levels with a decreasing number, until reaching a single objective at the top of the pyramid, with a view to define the preferability degree of the alternatives (considered in an indefinite number). This pathway involves the sharing of the "level of importance" to assign to the criteria

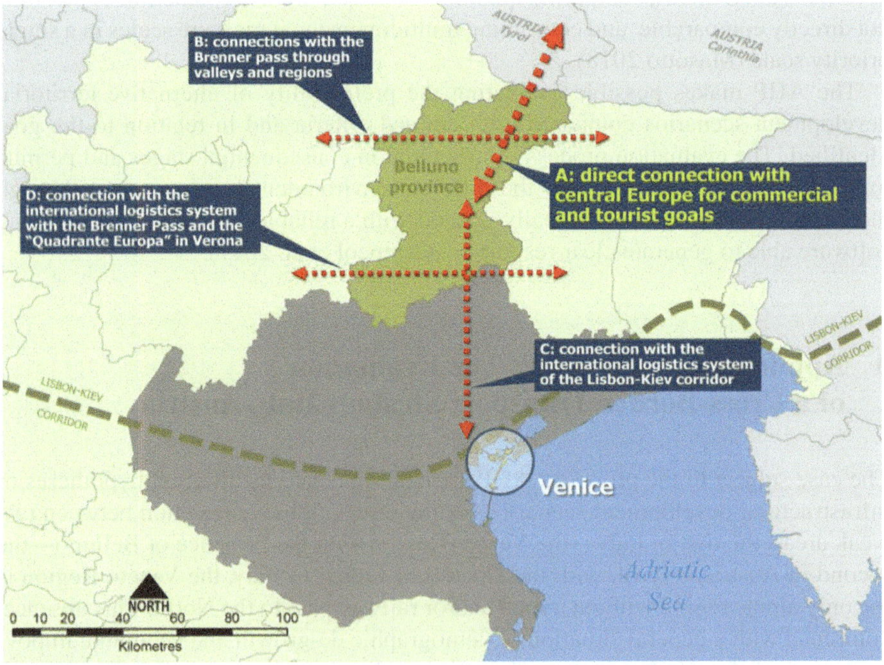

Fig. 1 Representation of the corridor Italy-Austria

within the WT, on the basis of the information obtained from the subjects involved, thus realizing a participatory moment—albeit on a small scale—(Masotto 2018).

The AHP is very helpful to solve a wide range of environmental problems and can help choose more sustainable solutions. The reliability of the results obtained can increase, while collecting the data and structuring the hierarchy of the AHP criteria, by using specific modalities, essentially specialized, that represent a theoretical basis to identify objective criteria for the weight allocation (Masotto 2018)

4 Discussion of the Results

The application of the AHP methodology has enabled laying down a dashboard, a kind of a "control panel" equipped with criteria and sub-criteria that have been assigned a weight (value), i.e. a tool permitting to visualize at a glance the most important information necessary to reach one or more goals (Few 2006). Here below are presented the fundamental aspects that may help to realize the dashboard, whose function is to support the decision-maker and also to direct future planning interventions (Masotto 2018).

In order to realize and apply the AHP model it is fundamental to start up a WT, through the preliminary identification of the participant categories close to the environmental issues identified, interested in participating as important stakeholders. Table 1 schematizes the competences required, and defines the role, interest and background of the participants (Masotto 2018).

Table 1 Diagram for the identification of the WT participants

CRITERI DI TIPO AMBIENTALE							CRITERI DI TIPO PROGETTUALE	
ASPETTI ABIOTICI	ASPETTI BIOTICI	ASPETTI UMANI						
ARIA, ACQUA, SUOLO	FLORA, FAUNA, HABITAT	SOCIETA'	PAESAGGIO	ECONOMIA		CONNETTIVITA' TERRITORIALE	VELOCITA' DI COLLEGAMENTO	COSTI DI INTERVENTO
				ECONOMIA TERRITORIALE	VALUTAZIONI IN CAMPO CIVILE, FORESTALE E AMBIENTALE			
- Interferenza sulla qualità dell'aria; - Interferenza sulla qualità delle acque superficiali; - Interferenza sulla struttura geologica e idrogeologica; - …	- Interferenza sulla flora, fauna e Habitat; - Interferenza sulle aree SIC e ZPS; - …	- Dinamica sociale; - Dinamiche economiche; - Dinamiche demografiche; - …	- Percezione dei paesaggi dolomitici; - Fruizione dei paesaggi dolomitici; - Interferenza su coni ottici storicizzati; - Interferenza sui coni ottici dei paesaggi del sito UNESCO delle Dolomiti; - Interferenza sui paesaggi degli insediamenti umani caratteristici; - …	- Sviluppo turistico; - Sviluppo del manifatturiero; - Sviluppo di nuovi servizi; - Sviluppo di centri di alta formazione; - …	- Stima costi di esproprio; - Stima perdita produttività agro-silvo-pastorale; - Sottrazione di aree boscate; - Sottrazione di aree prative; - …	- Generazione di nuovo ammagliamento infrastrutturale (realizzazione di opere per il miglioramento della viabilità locale); - Accesso a nuovi servizi (es. servizio sanitario, larga banda, ecc.); - Antropizzazione (fermare lo spopolamento); - …	- Velocità di comunicazione tra rete globale e la rete locale; - Velocità di comunicazione tra reti locali; - Velocità di fruizione dei servizi; - …	- Costi di progettazione; - Costi di costruzione; - Costi di gestione; - Project Financing; - Tempi di realizzazione; - …
Geologo, ingegnere ambientale, rappresentante politico	Naturalista, Forestale	Statistico, giurista, rappresentante associazione culturale, rappresentante fondazione culturale di studi e ricerche, rappresentante centro di solidarietà, consigliere comunale di Belluno	Valutatore paesaggio, architetto del paesaggio	Economista, rappresentante di categoria, rappresentante Ente Portuale di Venezia, rappresentante C.A.I., rappresentante politico	Valutatore, estimatore	Urbanista, rappresentante politico	Trasportista, progettista	Progettista, impresa costruttrice
13. … 22. … 32. …	17. … 14. …	11. … 12. … 24. … 27. … 29. … 30. …	04. … 15. …	02. … 10. … 23. … 25. … 26. … 28. … 31. …	03. … 05. … 08. …	21. … 33. … 34. …	01. … 09. … 18. … 16. …	06. … 07. … 19. … 20. …

Therefore, the participants ready to enter the WT were identified, and the WT was structured as follows:

- WT Coordinators
    - Graduate student;
    - Co-supervisor professor of the graduate student
    - Professor expert in the AHP methodology, from another university;
- University Experts
    - Some professors of the Academic Board for doctoral students
    - Other professors of other universities;
- Technical Experts
    - Engineer, former senior official of the State railways;
    - Planners of transport infrastructures (railways and roads);
- Stakeholders
    - Representatives of some cultural associations in the Belluno Province;
    - Representatives of some local traders' associations in the Belluno Province (hotels, tourism, industry, crafts, farming and forestry);
- Politicians
- Councilors of some Municipalities in the Province of Belluno
- Councilors of the Province of Belluno.

Six WT meetings were organized, followed by a final meeting for presenting the results, in which thirty-four participants took part. Each WT meeting positively contributed to the progress of the model construction (Masotto 2018).

It is possible to define two AHP evaluation models: "relative" (with a finite number of alternatives: AHP-R) and "absolute" (in the absence of alternatives or in the presence of indefinite alternatives: AHP-A).

The former needs to identify the main goal, of criteria and sub-criteria, by making absolute series comparisons and by identifying the ranking made up of a finite number of alternatives (measurable through an AHP "relative" model: AHP-R), through the adoption of a hierarchy organized in four levels (Masotto 2018):

- 1st level → evaluation goal;
- 2nd level → criteria considered for the evaluation;
- 3rd level → sub-criteria (specification of criteria);
- 4th level → alternative scenarios.

In this case, the definition of the existing alternatives as "indefinite" (since they cannot be compared due to their substantial planning heterogeneity) has in fact caused the elimination of the fourth level in the evaluation model (concerning alternative scenarios). Consequently, the decision-making structure has been organized in clusters on three levels, as shown in Fig. 2 (Masotto 2018).

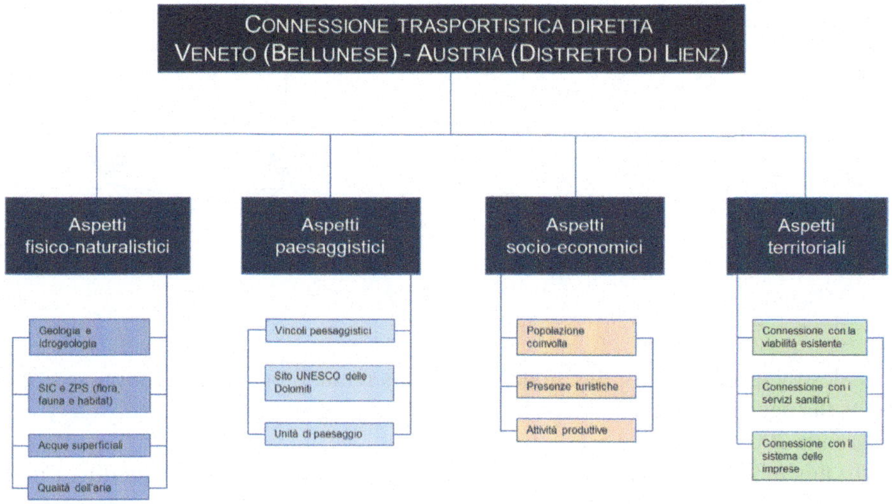

Fig. 2 AHP "absolute" conceptual model

- 1st level → evaluation goal: the direct road connection between Veneto Region (Belluno Province) and Austria (Lienz District);
- 2nd level → environmental criteria;
- 3rd level → specific sub-criteria, connected to an explanatory of the criteria.

The evaluation model envisages the elaboration of a questionnaire to be submitted to the WT. As shown in literature, it is important that all WT participants be favorable and agree on the pursuit of the main goal of the AHP model elaborated. In the experimentation presented, the WT produced different questionnaire versions, and each of them was tested by the same WT by first asking the participants whether they were in favor (yes/no) of the attainment of the main goal established in the model. While in the first two versions of the questionnaire there was not a full sharing of the goal, in the definite version the WT could refocus the main goal and identify the relevant criteria and sub-criteria, thus providing an added value to the experimentation (Masotto 2018).

5 Work Stages

The elaboration of the AHP-A model sets out essentially five work stages:

- Development of the hierarchy. In this first stage "[…] the decision-maker analyzes all the aspects of the problem and structures it in a hierarchy composed of several levels […]" (Caramia 2009). It is therefore a question of building the dominance hierarchy, "[…] a structural web composed of two or more levels […]" (Marchi and Lenti 2003) (Fig. 3).

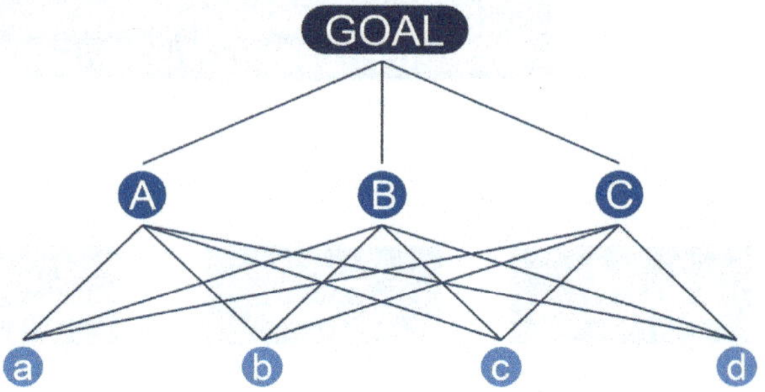

Fig. 3 Diagram of dominance hierarchy (Saaty 1980)

The destructuralization of the model permits its remarkable simplification and "enables the decision-maker to focus his analysis on a small number of decisions [...]" (Caramia 2009).

In this model the first level, i.e. the main goal, is represented by the direct road connection Veneto-Austria. This goal is evaluated through a set of environmental criteria linked to one another (second level), in their turn further defined by subcriteria (third level) and subject to pairwise comparisons until reaching a degree of importance (weight) of such criteria.

The alternatives to evaluate (only present in the AHP-R model) are placed at the bottom of the hierarchy and are directly connected to the most specific objectives (Giangrande 2008).

- Construction of the pairwise comparison matrix: "[...] it consists in identifying the estimated "weights" to match with each criterion that exist in the hierarchical problem by using an evaluation matrix whose individual elements are obtained from the pairwise comparisons of the problem criteria [...]" (Caramia 2009). The structure of this matrix is reported in Table 2:

[...] The elements of each pair are compared in order to establish which of them is more important, in relation to a super-ordered element, and to what extent: the

Table 2 Structure of the pairwise comparison matrix (Saaty 1980)

n	1	2	3	...	m
1	1		valori reciproci		
2		1			
3	giudizi		1		
...				1	
m					1

result of the comparison is the dominance coefficient aij; which represents an evaluation of the "dominance" of the first element (i) compared with the second (j) […]. (Giangrande 2008).

- Determination of the relative local weights: for all the pairwise comparisons matrices, the "weights" of all elements of the hierarchy are matched to each criterion. These "weights" are considered "local" since they evaluate the importance of the elements not as a whole but only in relation to the super-ordered element they have been compared with. Each element has as many "local weights" as the objectives it is directly subordinated to (Giangrande 2008).

The "local weights" are therefore determined through the pairwise comparison, and the quantification of the relative importance of the different criteria arises from the preference declaration in relation to the use of the "Saaty's semantic scale" with values 1 to 9 (Campeol et al. 2016) (Table 3), that "correlates the first nine entire numbers with as many judgments that qualitatively express the possible results of the comparison […]" (Saaty 1980; Giangrande 2008). The "weight" allocation makes it possible to compare different judgments, thus producing a synthetic and comprehensive evaluation of the territorial performances.

- Analysis of the judgment consistency: "[…] In this stage it is necessary to verify whether the "weights" obtained in the previous stage are faithful to the judgments expressed by the expert […]" (Caramia 2009). Since the expert does not have an objective measurement tool but rather he uses his own technical and scientific skills (know-how), he is not able to directly determine the "weights". As a consequence, he can only give approximate estimates of their relations with the help of Saaty's scale. Therefore, in most cases the estimates provided by the expert cannot be consistent, due both to the difficulties that the expert meets in keeping a consistent judgment in all pairwise comparisons, and to the fact that his judgments can structurally be inconsistent (Giangrande 2008).

Table 3 Semantic scale of Saaty (1980)

Intensity	Definition	Explanation
1	Equal importance	Two activities contribute equally to the objective
3	Moderate importance	Experience and judgement slightly favour one activity over another
5	Strong importance	Experience and judgement strongly favour one activity over another
7	Very strong importance	An activity is favoured very strongly over another; its dominance demonstrated in practice
9	Extreme importance	The evidence favouring one activity over another is of the highest possible order of affirmation
2, 4, 6, 8	Intermediate values	When a compromise is necessary

To this aim, the AHP method defines a consistency index (CI), "[…] that enables measuring the total gap between these two sets of values […]" (Giangrande 2008). Therefore, the derived matrix can be analyzed by using the consistency index, which makes it possible to evaluate to what extent the "weights" derived are consistent with the decision-making process.

- Determination of the global weights (principle of hierarchical composition): the calculation of the "global weights" (or priorities) of the alternatives represents the culmination of the evaluation process. This calculation is performed by multiplying the "local weights" of each element by those of the corresponding superordered elements, and then summing up the products obtained. "To determine the importance of each element in relation to the goal, it is necessary to apply the principle of hierarchical composition […]" (Saaty 1980; Giangrande 2008). With a top-down approach, the "local weights" of all the elements in the hierarchy are therefore progressively transformed into "global weights" […] (Giangrande 2008).

In a four-level hierarchy (goals, criteria, sub-criteria and alternatives) typical of an AHP-R model, the "global weights" (or priorities) of the elements, placed at the bottom of the hierarchy (in the level following that of terminal objectives), represent the main result of the evaluation. When the terminal elements are the alternatives (Giangrande 2008; Caramia 2009), "[…] the "global weights" permit to determine an order of preference […]" (Giangrande 2008; Caramia 2009): therefore, the more "global weight" an alternative (a scenario) has, the more preferable it is […]" (Giangrande 2008; Caramia 2009).

6 Outcomes

The AHP application devises a questionnaire to be submitted to the Working Table participants, in which they are asked to express their judgment through pairwise comparisons, especially between the four identified criteria and then between each set of sub-criteria (Fig. 4). The number of the questions has been defined on the basis of the rule of reciprocity (Saaty 1980) and, in the specific case, the result is equal to 21: six pairwise comparisons for the 2nd level (a 4×4 matrix), 15 pairwise comparisons for the 3rd level (one 4×4 matrix, three 3×3 matrices).

The seventeen participants have been grouped in four categories:

- Four University Experts (UE);
- Four Technical Experts (TE);
- Five Stakeholders (SH);
- Four Politicians (PO).

The elaboration of the replies has been carried out with an open source computing software. The below graphs (Figs. 5 and 6) summarize the judgments expressed by the four groups of participants (UE, TE, SH, PO) in relation to the criteria (Physical

1. Criteri di confronto nel secondo livello gerarchico

1.1 Dal suo punto di vista, quale criterio è più importante per valutare un'ipotesi di connessione trasportistica (ferroviaria/stradale) diretta Veneto (Bellunese) - Austria (Distretto di Lienz):

FN: ASPETTI FISICO-NATURALISTICI
PA: ASPETTI PAESAGGISTICI
Quale criterio è più importante? FN PA
In quale misura? 1 3 5 7 9

FN: ASPETTI FISICO-NATURALISTICI
SE: ASPETTI SOCIO-ECONOMICI
Quale criterio è più importante? FN SE
In quale misura? 1 3 5 7 9

FN: ASPETTI FISICO-NATURALISTICI
TE: ASPETTI TERRITORIALI
Quale criterio è più importante? FN TE
In quale misura? 1 3 5 7 9

PA: ASPETTI PAESAGGISTICI
SE: ASPETTI SOCIO-ECONOMICI
Quale criterio è più importante? PA SE
In quale misura? 1 3 5 7 9

PA: ASPETTI PAESAGGISTICI
TE: ASPETTI TERRITORIALI
Quale criterio è più importante? PA TE
In quale misura? 1 3 5 7 9

SE: ASPETTI SOCIO-ECONOMICI
TE: ASPETTI TERRITORIALI
Quale criterio è più importante? SE TE
In quale misura? 1 3 5 7 9

2.2 Dal suo punto di vista, quale criterio è più importante, tra gli aspetti paesaggistici qui considerati, nel valutare un'ipotesi di connessione trasportistica (ferroviaria/stradale) diretta Veneto (Bellunese) - Austria (Distretto di Lienz):

PA1: Vincoli paesaggistici
PA2: Sito UNESCO delle Dolomiti
Quale criterio è più importante? PA1 PA2
In quale misura? 1 3 5 7 9

PA1: Vincoli paesaggistici
PA3: Unità di paesaggio
Quale criterio è più importante? PA1 PA3
In quale misura? 1 3 5 7 9

PA2: Sito UNESCO delle Dolomiti
PA3: Unità di paesaggio
Quale criterio è più importante? PA2 PA3
In quale misura? 1 3 5 7 9

2.3 Dal suo punto di vista, quale criterio è più importante, tra gli aspetti socio-economici qui considerati, nel valutare un'ipotesi di connessione trasportistica (ferroviaria/stradale) diretta Veneto (Bellunese) - Austria (Distretto di Lienz):

SE1: Popolazione coinvolta
SE2: Presenze turistiche
Quale criterio è più importante? SE1 SE2
In quale misura? 1 3 5 7 9

SE1: Popolazione coinvolta
SE3: Attività produttive
Quale criterio è più importante? SE1 SE3
In quale misura? 1 3 5 7 9

SE2: Presenze turistiche
SE3: Attività produttive
Quale criterio è più importante? SE2 SE3
In quale misura? 1 3 5 7 9

2. Criteri di confronto nel terzo livello gerarchico

2.1 Dal suo punto di vista, quale criterio è più importante, tra gli aspetti fisico-naturalistici qui considerati, nel valutare un'ipotesi di connessione trasportistica (ferroviaria/stradale) diretta Veneto (Bellunese) - Austria (Distretto di Lienz):

FN1: Geologia e idrogeologia
FN2: SIC e ZPS (flora, fauna e habitat)
Quale criterio è più importante? FN1 FN2
In quale misura? 1 3 5 7 9

FN1: Geologia e idrogeologia
FN3: Acque superficiali
Quale criterio è più importante? FN1 FN3
In quale misura? 1 3 5 7 9

FN1: Geologia e idrogeologia
FN4: Qualità dell'aria
Quale criterio è più importante? FN1 FN4
In quale misura? 1 3 5 7 9

FN2: SIC e ZPS (flora, fauna e habitat)
FN3: Acque superficiali
Quale criterio è più importante? FN2 FN3
In quale misura? 1 3 5 7 9

FN2: SIC e ZPS (flora, fauna e habitat)
FN4: Qualità dell'aria
Quale criterio è più importante? FN2 FN4
In quale misura? 1 3 5 7 9

FN3: Acque superficiali
FN4: Qualità dell'aria
Quale criterio è più importante? FN3 FN4
In quale misura? 1 3 5 7 9

2.4 Dal suo punto di vista, quale criterio è più importante, tra gli aspetti territoriali qui considerati, nel valutare un'ipotesi di connessione trasportistica (ferroviaria/stradale) diretta Veneto (Bellunese) - Austria (Distretto di Lienz):

TE1: Connessione con la viabilità esistente
TE2: Connessione con i servizi sanitari
Quale criterio è più importante? TE1 TE2
In quale misura? 1 3 5 7 9

TE1: Connessione con la viabilità esistente
TE3: Connessione con il sistema delle imprese
Quale criterio è più importante? TE1 TE3
In quale misura? 1 3 5 7 9

TE2: Connessione con i servizi sanitari
TE3: Connessione con il sistema delle imprese
Quale criterio è più importante? TE2 TE3
In quale misura? 1 3 5 7 9

Fig. 4 Questionnaire

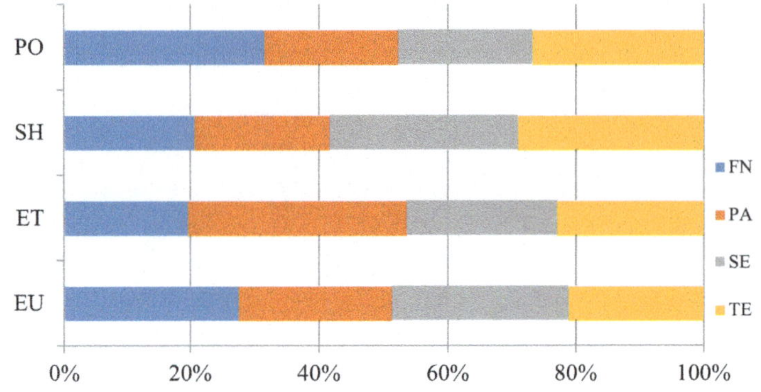

Fig. 5 Criteria weighing

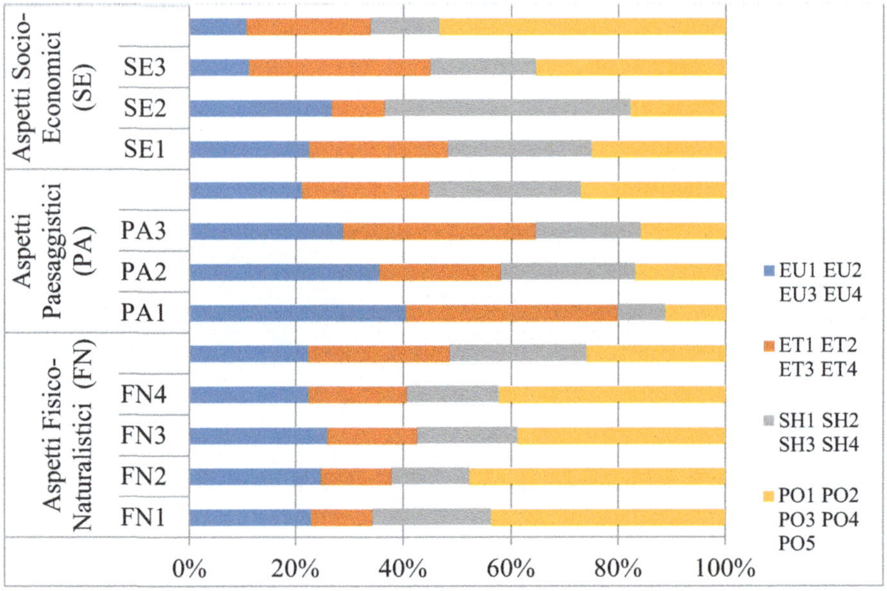

Fig. 6 Sub-criteria weighing

and Natural aspects PN, Landscapes LA, Socio-Economical SE and Territorial TE) and to the sub-criteria, expressed in percentage values.

The graph in Fig. 4 highlights, in short, that the Technical Experts consider the landscape aspects as prevalently important (36%), the Politicians refer to the physical and natural aspects (35%) followed by the territorial aspects (30%), the Stakeholders underline the importance of the socio-economic aspects (26%); on the other hand, the University Experts consider the four aspects as essentially equivalent.

Moreover, it is remarkable to observe that the socio-economic aspects are considered equally important for all groups of participants. This confirms the importance of interventions aimed at supporting local, touristic and manufacturing development.

The graph in Fig. 5 shows, when weighing sub-criteria on the physical and natural aspects, the primary importance attributed by Politicians to the SCI and SPA areas, followed by geologic and hydro-geologic problems, air quality, and finally by the physical characteristics of water courses in general and of those subject to erosion. However, it highlights a discrepancy in the judgments expressed by the Politicians between the considerable attention given to the connection with enterprises and a lower importance of the connection with the existing roads, which is instead considered more important by the Stakeholders.

Moreover, there is homogeneity for the physical and natural aspects also for the University Experts, who consider them as important as the socio-economic ones and as the core zones and buffer zones (part of the UNESCO site of the Dolomites), followed by the characteristics of the various landscape unities.

Nevertheless, these representations, arising from the elaboration of the preference system expressed by the participants, should not be interpreted tout-court as a global preference system; rather, they represent a verification of the consistency between the information provided and the results obtained.

7 Considerations

The evaluation of the environmental sustainability of a cross-border transport development program prioritizes the SEA. It is important to underline that the ex-ante stage of the SEA, strategically important since it permits to set out environmental potentialities and criticalities before defining transformation scenarios, is less developed in terms of methodology and application. The AHP model experimented, correctly set within the SEA process, is used in the ex-ante stage for the definition of the strategic scenarios of infrastructural development (i.e. the stage with the highest capacity of identifying strategic visions). Therefore, this experimentation represents a methodological progress in the framework of the evaluation itself (Masotto 2018).

Moreover, it must be noted that in the elaboration of the AHP-A evaluation model and its application to the case study, the WT has played a crucial role since it has enabled the refocusing of the main goal and the identification of criteria and sub-criteria, thus providing an added value to the experimentation (Masotto 2018).

Thanks to its versatility and flexibility, the AHP can be adapted and used for different types of problems, in terms of environmental assessment, by experts and professionals, but not just those. Its structure, in relation to the latest evolution of the Analytic Network Process (ANP), presents in the SEA a high effectiveness because of its higher capacity of simplification. In fact, it should be remembered that an evaluation method is effective if it is capable to simplify the environmental and planning complexity.

The experimentation carried out has shown that the AHP can make the SEA more effective, bringing together and organizing an environmental reference framework with modalities that can be used by experts called upon to express their "preferences".

References

Campeol G, Carollo S (2003) Nuovo Piano Urbanistico di Ferrara: VALSAT Preliminare Quadro Conoscitivo. Iuav Studi & Progetti, Venice

Campeol G, Carollo S, Masotto N (2016) Progetti Infrastrutturali e Valutazione Ambientale per uno Sviluppo del Territorio Dolomitico Bellunese. In: XXXVII Scientific Conference AISRe Quali confini? Territori tra identità e integrazione internazionale, 20–22 September. Ancona

Campeol G, Carollo S, Masotto N (2017) Development theories and infrastructural planning: the Belluno province. In: Bisello A, Vettorato D, Stephens R, Elisei P (eds) Smart and sustainable planning for cities and regions. Results of SSPCR 2015. series green energy and technology. Springer International Publishing. Cham (CH), pp 299–315. https://doi.org/10.1007/978-3-319-44899-2_18

Caramia M (2009) Analisi Multiobiettivo Multicriteria. Course of Progettazione e Simulazione dei Sistemi di Produzione e di Servizio. Department of Business Engineering Mario Lucertini. University Tor Vergata. Rome

Few S (2006) Information dashboard design: the effective visual communication of data. O'Reilly Media, Sebastopol USA

Giangrande A (2008) AHP (Analytic Hierarchy Process): Teoria ed esempi di valutazione di progetti alla scala territoriale. Course on Tecniche di valutazione ambientale di piani e progetti. Roma Tre University. Rome

Golden BL, Wasil EA, Levy DE (1989) Application of the analytic hierarchy process: a categorized, annotated bibliography. In: Golden BL, Wasil EA, Harker PT (eds) The analytic hierarchy process. Applications and studies. Springer, Berlin

Masotto N (2018) Valutazione ambientale nelle trasformazioni territoriali in ambito alpino: applicazione del metodo AHP, doctoral thesis in Engineering Management and Valuation, specialization in Estimate and Regional Economies—Cycle XXIX

Saaty TL (1980) The analytic hierarchy process. McGraw-Hill Book Co., New York

Saaty TL (1986) Axiomatic foundation of the analytic hierarchy process. Manage Sci 32:841–855

Urban Strategies in the Legislation of the Veneto Region and the SEA

Vincenzo Fabris

Abstract Traditional urban planning is unable to carry out its tasks. The ongoing economic and territorial trends are phenomena that were unthinkable and unforeseeable some decades ago. The "plan" and the planning system are therefore inadequate and it is necessary to reconsider the discipline and its tools. The disciplinary "reorganization" is not accomplished yet, in spite of the multiple innovation attempts. The new urban planning shall be based on the innovations made available by technology, on reliable and appropriate informative systems, and it will be necessary to adopt evaluation methods derived from instruments such as the SEA, the EIA, the AIES, etc. The goals are inverted: the assumption shall be assessed, and not the plan that makes the hypothesis. The plan lies in the method and in the evaluation system, not in the prefiguration of an urban project.

Keywords Urban planning · SEA · Veneto region · Cognitive framework

For understanding the ongoing dynamics and trends in the urban planning of the Veneto Region, it is necessary to shortly trace back the history of regional planning.

The founding act is the Italian National Law n. 1150 of 1942, which gave a regulatory and disciplinary framework to all subsequent legislative provisions,[1] also at national level.

It is good to remember that the first urban laws of the Veneto Region were Laws n. 58 of 1978 (protection and buildings in rural areas) and n. 73 of the same year (extension of productive activities in improper areas). In other words, the first concern of the legislator was to recognize a territorial specificity that had not duly been considered in the national legislation: in 1978 there was already a problem of settlement dispersion that Law n. 1150/42 had neither understood nor addressed in terms of planning.

[1]Italian Laws n. 167/62, n. 765/67, n. 865/71, n. 457/78, and then the Bucalossi and Nicolazzi Laws, Ministerial Decree 1444/68 and so on.

V. Fabris (✉)
Urban Planning Office in the Veneto Region, Venice, Italy
e-mail: vincenzo.fabris@alice.it

G. Campeol (ed.), *Strategic Environmental Assessment and Urban Planning*, Green Energy and Technology, https://doi.org/10.1007/978-3-030-46180-5_6

Considering the above, it must be said that the present regional planning system is a direct consequence of the crisis of the second half of the 80 s. In that period the Urban Plan, or better, the General Urban Development Plan and all the theoretical planning construction had reached a limit of practical unsustainability, thus dramatically showing all the problems of the system.

These problems can be summarized as follows:

- Lack of flexibility
- Cumbersome procedures
- Excessive planning details
- Excessive regulations (difficulties to interpret the tools and so to manage the territory)
- Excessive planning (an infinite number of plans, often for the same matter)
- Never-ending disputes.

Inconsistency with the territorial situation (see, for example, the urban standards that were surely at the forefront in 1968, but that today no longer meet the present social needs).

A situation of extremely difficult manageability had been generated, with plans accompanied by more than 14,000 comments and 1,500 pending cases. For each problem, although small, the Plan had to include variants: some provinces had between 300 and 400 variants to be approved each year. Old plans were characterized by inertia, i.e. the impossibility to lay down new plans without recognizing previous situations. The costs of plan elaborations were rocketing, and disciplinary development had become almost self-referential (Byzantine city planning).

All this led also the Veneto Region to recognize the need for a new framework law on land planning.[2]

At the end of the 90s a process was started, leading to the issuing of Italian Law n. 11 of 2004 (to tell the truth, already in 1993/1994 the first attempts to produce a new legislation, although unfruitful, were made).

Undoubtedly, the cultural and disciplinary reference was to the theses of the INU (National Institute for Land Planning) expressed in the 1995 congress, especially the adoption of a Structural Plan/Operational Plan, instead of the PRG (General Urban Development Plan).

The new Italian Law (n. 11 of 2004) sets out the division of the Plan into the Land Use Plan (strategy Plan) and Intervention Plan (operation Plan). The former is approved by the higher authority and the latter by the municipality.

The pyramid architecture of the planning system (large plan-medium plan-small plan, i.e. PTRC/PTCP/PAT) is maintained, although with some important flexibility measures: the higher plan determines the modification framework allowed to the subordinated tool.

There are also important technical innovations.

Here below they are randomly listed:

[2]The first real Land Planning Law of the Veneto Region is n. 40 of 1980 whereas in 1985, Law n. 61 was approved to essentially adjust Law n. 40.

- Institution of the format template and of the index of data quality
- Variability of urban standards
- Unification of implementation tools
- Estimate of volumetric credits
- Institution of regional strategic projects
- Institution of urban equalization and compensation
- Term (5 years) for the territorial transformation projections, if unimplemented
- Admissibility of public/private agreements
- Concentration as a method of forming urban planning tools
- New list of public services
- Landscape discipline
- SEA.

As here shown, this is a huge number of innovations.

However, the most important innovation is the different approach to planning. It is no longer a land planning law, but rather a law on the territory government.

The goal is not just to plan the city, but also to govern the rules of the territory use, in its broadest definition.

This foreshadows a new legislation on the "House Plan", land use, and urban regeneration.

It is interesting to note as in art. 2 (Content and Purpose) the Italian Law n. 11 clarifies that it is allowed "… to use new territorial resources, only when no regeneration alternatives exist….".

Moreover, the projection of a maximum amount of a consumable agricultural area and the definition of the "format template" are introduced.

This last institute completely reverses the concept of territorial analysis.

Until 2004, the disciplinary reference was composed of the regional unified graphic representation of 1983, based on the sequence analysis/project/verification. Information was collected, plans were made, and finally the project consistency with technical specifications was verified.

The problem was the cumbersome system (and also its cost): the collection of a huge amount of data was often unnecessary to the project elaboration.

The idea underlying the format template is radically different.

The prerequisite is the existence of a territorial information system that is today an indispensable tool for administering any Plan management choice, which enables information updates and constant verification of the Plan outcomes.

The sequence becomes project/verification/monitoring.

An elaboration method, similar to the evaluation techniques of the SEA, EIA, and AIES is now used for all the sectorial planning (PAI, etc.).

Finally, the SEA is another extremely important new element of Italian Law n. 11.

It had been discussed for long, when the elaboration of the new regional land planning law was started. However, Directive n. 42 of 2001 had not been transposed into the Italian legislation yet, so that Art. 4 of Law n. 11 was affected by the interpretative uncertainty that was reduced only with the successive measures, and therefore everything was left to general guidelines: transposition in 2006 with effectiveness in 2007. In that moment, it was considered that the SEA was a planning internal measure, and also the only way to come to a planning decision.

The discipline has obviously evolved, and new concepts and terminologies have developed in urban planning.

The SEA is not the system for elaborating either the land planning schemes or the IPs, or even the implementation plans.

It is a procedure that is added to the Plan and verifies its sustainability depending on the environmental variables.

However, the concepts and the basics of the SEA, first of all the attention to the environment, have become an essential day-to-day component in planning, in primis the concept of "impact assessment".

Planning had never been used to measure the possible consequences of urban planning projections.

Similarly, the analysis of alternatives has never been a support to the Urban Development Plan. As regards monitoring, it has been considered a constant control of the effects, but also a feedback of the attainment of expected results (both in the "House Plan" and in land use).

Through the environmental assessment, the use of indicators, the prospect of compensatory measures, mitigation actions, etc. are modalities that are increasingly developing in the planning process.

The analysis of the zero option is worth emphasizing. In fact, in this very difficult economic situation, more and more frequently the zero volume Plans are evoked. Stop to territorial expansions consuming the soil, etc.

In fact, the signs of a radical ongoing change are evident; this is not only a crisis, but also a completely new scenario, rapidly evolving, whose outcomes cannot be envisaged yet. There is a debate on the need to consider new transformations, when the demand comes to an end, in front of an availability valued at approximately 80 to 100 million m3, on a regional basis. Flattening the Plan on the zero option, paradoxically, may not be a virtuous operation; instead, it can be a dangerous drift against the ongoing evolution.

It is good not to consider new transformations, but this means overlooking the landscape, the historical centers, the suburbs, the soil protection, the quality of settlements, the new needs, the elderly, accessibility, safety, etc.

In other words, it is necessary to think about a planning action that should aim at the territorial—and only afterwards, urban—regeneration, developing from a zero option as a starting point and not as a goal to be achieved.

In this context, the question of information becomes of crucial importance. The territorial dynamics arise and develop much more quickly than ever before in terms of planning. The plans had, and still have, unlimited duration, to the extent that the projections included, although evidently obsolete, lasted for decades.

These dynamics were tumultuous but, especially, always evolving at a very high speed. The traditional Plan, which imagines and prefigures the city, is inappropriate and is no longer effective because it is static and crystalized in postulated choices, with very low adaptability.

The format template outlined by Italian Law n. 11 can represent an interesting working hypothesis.

Art. 10 of Italian Law n. 11 states that "the format template is the integrated system of the necessary information and data for understanding the issues developed by the territorial and urban planning tools".

The format template is part of the TIS (Territorial Information System) and it is the basis for every planning tool.

The 2010 Guidelines define it as a "Plan report, but at the same time as an autonomous tool".

A contradiction emerges: it is surely an autonomous set of information, that can be autonomously upgraded, modified and certified, but the chronological representation is part of the Plan and justifies its choices.

It is important to underline that the version of the 2010 Guidelines in also linked (as expressly stated) to the entry into force of the new SEA regulations.

The format template is made up of 341 coded classes grouped in folders. Most of these 11 folders concern environmental issues.

- C 01, Territorial information (map)
- C 02, air
- C 03, climate
- C 04, water
- C 05, subsoil
- C 06, biodiversity
- C 07, landscape
- C 08, architectural heritage
- C 09, physical invariants
- C 10, economics and society
- C 11, planning and constraints.

It can be stated that most analyses and information, acquired with the PAT (Land Use Plan) and the other planning tools, concern issues referred to the SEA.

It should be noted that a revision, also methodological, of the format template has been carried out for long, and the matter involves directly the SEA and the methods of territorial analysis.

There are lots of questions to answer:

- What is the reliability of information?
- Can a certification value be attributed?
- What is the level of information updating?
- What is the propensity of interaction with other databases?
- What is the level of accessibility and usability?
- What are the updating methods?

All these evaluations are prior to each analysis of the format template, which in turn is prior to each Plan.

Very briefly, without a correct format template there is neither a correct (well informed) SEA nor a Plan.

In conclusion, the priority is to be fully aware that the territorial scenario of urban planning has radically changed.

The Plan is no longer—and it will never be—the arena where all territorial dynamics are measured and balanced.

On the contrary, the trend that is increasingly developing is to avoid any planning, as shown by laws, institutes, provisions that make it possible to put aside urban planning.

The planning process should probably be reversed. It should no longer be a Plan that the government action must comply with; rather, it should verify the compatibility of the government action with the information system.

If this were the case, the format template would become crucially central, both for the Plan and for the evaluation system.

It could be stated that when it is impossible to develop a global view of the government action on the territory, the availability of a correct and reliable format template may be sufficient to ensure good planning.

Indeed, planning is the format template.

References

AA. VV (1983) Grafia e simbologia regionali unificate per la redazione degli strumenti urbanistici. Regione del Veneto. Venezia

Artico L et al (1987) Legislazione regionale in materia di territorio. Regione del Veneto, Venezia

Barel B (a cura di) (2004) La Legge urbanistica della Regione Veneto. S.I. Corriere del Veneto, Milano

Barel B (a cura di) (2017) Contenimento del consumo del suolo e rigenerazione urbana. Atti del Convegno Consumo di suolo e rigenerazione urbana nella nuova legge veneta, Vicenza

Micelli E (2004) Perequazione urbanistica. Marsilio, Venezia

Turri E (2000) La megalopoli padana. Marsilio, Venezia

The Effectiveness of SEA in Evaluating the Land Consumption. An Analysis on 100 SEAs in Lombardy, Italy

Rossella Moscarelli and Paolo Pileri

Abstract SEA is a tool useful to evaluate the environmental effects on the territory, generated by the urban transformations. In this study, the land consumption is the effect we take into account. Indeed, the paper aims at investigating how SEA is effective in evaluating it and, in case, in reducing the land consumption generated by the urban plans at municipal scale. The analysis considers the Italian context, in particular the Lombardy Region, which was, in 2005, the first Region in Italy having implemented the European Directive on SEA. Assuming that in Italy the land transformations are mostly decided by the municipal planning, 100 SEAs related to 100 different municipal urban plans, have been studied. It is the widest sample of SEA ever studied until now in Italy. The purpose of this research consists of verify how usually SEAs treat the issue of the land consumption and if they provide indicators, uniform and comparable each other, to measure it. First, it is introduced how SEA can evaluate the land&soil issue, starting from the initial intentions of the European Directive of 2001, to the translation in the Italian legislation. Then, it is explained the specific case of the Lombardy Region (in Italy the authority about these topics are transferred to each Regions), with its recent urban planning laws, the regional laws, RL 12/2005 and the RL 31/2014. The former changed the planning system at municipal scale; the latter wanted to govern and reduce the land consumption derived from the urban plans. In the fouth section, it is presented and discussed the indicators introduced by the RL 31/2014 to calculate the land consumption. The following part shows the analysis of 100 SEAs, considering whether they evaluate the land&soil issue and how they do (using the new indicators provided by the RL 31/2014 or not). In the conclusion the results of the analysis are commented, considering that the new Regional law did not success in making uniform the approach to evaluate and measure the land consumption.

Keywords Land consumption · Lombardy region · Regional law · Urban plans

R. Moscarelli (✉) · P. Pileri
Department of Architecture and Urban Studies DASTU, Politecnico di Milano, Milan, Italy
e-mail: rossella.moscarelli@polimi.it

P. Pileri
e-mail: paolo.pileri@polimi.it

G. Campeol (ed.), *Strategic Environmental Assessment and Urban Planning*, Green Energy and Technology, https://doi.org/10.1007/978-3-030-46180-5_7

1 SEA and Soil in the Italian Context

SEA Directive has to be applied to a wide range of urban plans and programs. Regarding the land use issue in the planning field, there is no doubt that SEA could be effective just if it is inextricably linked to decision-making processes (Treville 2011). Indeed, urban plans modify the territory, potentially compromising a large number of environmental resources. Sometimes in an irreversible manner. To minimize and control the effects generated by the decisions of the urban plans, SEA, by law (SEA Directive, Annex 1, letter f), can/must:

- evaluate the main possible effects on the environmental resources, such as soil, water, air and landscape;
- integrate environmental considerations and evaluate the inter-linkage with economic and social aspects (soil as an ecological resource, soil as a 'common');
- include monitoring tools (e.g. about land consumption) to guide plan and to correct policies and actions on going.

The European Directive on SEA is in force since 2001. Each national legislation should transposed it by July 2004. Nevertheless, in Italy, SEA came into force only in April 2006, with the Legislative.

Decree n. 152. This delay was probably due to bureaucracy, worsened by a low ecologic culture and/or by a lack of awareness on the necessity and on the value of the environmental evaluation in urban planning. After the national acceptance, it was the turn of the Regions to make effective the law (in Italy urban planning legislation is transferred to each single Regions). The process of acceptance of the National Directive has been very long and it was characterized by two main periods. The first moment was from 2006 to 2012: each Region elaborated a law or some procedures on SEA to apply the national law (Sini 2016). The second moment lasts from 2013 to present: it refers to some upgrades and changes in the urban planning procedures so to better take into consideration SEA expectations. This second step is particularly important, considering the soil resource. Indeed, in Italy, soil is not officially regarded as an environmental resource and there is not a national law on this issue even if the topic of land consumption is emerging into the public discussion as a great problem affecting the future. Well, in the past, there were any urban planning laws defining what soil was (only in 2014 soil has been better defined[1]), despite the fact that main subject of the urban planning is the land, which is made mostly by soil, ant its transformations. Moreover, the ecological role of the soil has been even less defined. This is a key point in Italy, since the Constitution states that just the State is responsible of the ecological/ecosystemic resources, not the Regions. But, until no one say that the soil is an ecosystemic resource, as it is by facts, it will remain a 'strange' object, more similar to a platform useful to build than an alive, not renewable, ecological body. Therefore, being the urban planning the discipline that has treated more than the others the issue of the land&soil, the idea of soil as ecological resource has been substituted by the image of land. This confusion between soil and land, on the one

[1] We are referring to the Legislative Decree n. 46, 2014.

hand, certainly does not help the application of SEA procedure and, on the other hand, contributes in reducing the cultural acceptance of SEA by public technicians, professionals and politicians. The lack of an official and legal recognition of soil as an ecological resource raises a question about SEA's procedure, whether it should consider in its evaluation the soil and the land consumption or not. The answer is not easy, even if there are scientific evidences that prove the importance of the soil as environmental resource (AA.VV 2015). The resolution of this ambiguity depends on the importance given to the issue of land&soil in the each regional legislative frameworks and on the updating of the planning laws (and their goodness). Even if the theme of land&soil has been better treated over these years, some aspects still do not work well.

In addition to the legislative aspect, there are also two other problematic issues, that contribute in weaken SEA procedure. One is linked to the fact that topics as land&soil, being complex and related to changing laws, need to be well known by the various decision makers involved in the planning process and therefore the Regions should provide education and continuous updates. The second depends on the problem that each Region has elaborated different procedures in the application of SEA to urban plans, increasing entropy and relegating SEA in the field of the useless bureaucracy. Indeed, focusing to the use of land&soil, we are not able to register any significant improvement since SEA exists. From 2005 to present, in Italy, the land consumption has not decreased thanks to the application of SEA or, at least, no one has demonstrated this or tackled this topic. If the land consumption has declined, this is due to the crisis and to a little increase in popular and political awareness about the environmental issues.

2 The Case Study: The Lombardy Region

As explained above, analyzing urban planning tools in the Italian context implies the necessity to choose a specific Region, since each one differs from the others for what concern the legislation and the practices. For this research on SEAs, we have chosen the Lombardy Region, in the north of Italy. This selection was done for two main reasons. The former is that the Lombardy was the first Italian Region, which introduced SEA into its regulatory system, in 2005. It anticipated also the national Decree of the 2006. Since in Italy urban planning is a matter of exclusive regional jurisdiction, the Lombardy could incorporate SEA procedure into its new urban planning law (that will be deepen in the next paragraph).

The latter reason is related to the fact that Lombardy is the most economically strong Region in Italy, the most populous, with the highest growth demographic expectations. This also means that the environment is more subjected than other situations to pressures due to instances of landscape transformations, to environmental risks and land&soil consumptions. Not by chance, Lombardy is the strongest land consumer in Italy: with more than 12% of urbanized areas, Lombardy is the most sealed Region in Italy. In 2017, Lombardy has been the second region for annual increase of new urbanized areas (Munafò 2018). In such situation, it is important that protective tools work out efficiently.

In order to understand how SEA can operate in the context of specific urban planning tools, it is useful to deepen the role of the most relevant and recent laws elaborated by the Lombardy Region on this field. The first one it the RL n. 12, 2005, on the redefinition of the urban planning; the second one is the RL n. 31, 2014, on the soil and reduction of land consumption.

3 Urban Planning in Lombardy: The Regional Law 12/2005

In Lombardy, the urban planning system has been organized by three levels: regional, provincial and local (i.e. the municipal scale). Regional and provincial planning tools aim at giving a strategic vision, more than at designing the future transformations of the territory. It is the local planning that manages more than 90% of the transformations of land&soil. Since it depends on local initiatives, made by more than 1500 independent units, the Municipalities, the result is the collection of more than 1500 different ways of transforming land&soil. It is hard to control. But it becomes even harder if the urban planning law contributes in making things more complex than they can be. It happened with the RL 12/2005, when the Region defined the following scheme. The urban planning framework in Lombardy, at local level (municipal scale), has been made up by three sub-planning tools that work together on different levels:

- local strategic plan (*documento di piano*);
- local set of planning rules (*piano delle regole*);
- welfare&facilities urban plan (*piano dei servizi*).

The local strategic plan provides the general guidelines for urban planning and defines which areas can be transformed in the future. For each area to be transformed, a land use destination is assigned (urban, rural, natural), but the developing indicators are not specified. Generally, in this document it is not possible to find anything fixed about the amount of surface will be sealed (land consumed), nor how much building volume is permitted, the type of industrial activity will be localized, etc. These quantifications/specifications are assigned to the planning rules document. Finally, the facility plan deals with the dimensioning and localization of all public or private facilities. Even if these two last sub-plans have got sharply information for SEA procedure, while the local strategic plan does not have, Lombardy decided, in 2005, that only the local strategic plan could be submitted to SEA. On the contrary, the submission to SEA of the other two urban tools have depends on a preliminary verify of inclusion/exclusion made by the consultants and the Municipalities themselves. It was an unusual choice, but not criticized by anyone. A choice that has limited (and still limits) the task of SEA by relegating it to a sterile procedure. Indeed, it is very difficult to make an environmental assessment on determinants whose exact role is not known a priori. The risk is to underestimate or not to take into account the impacts that will instead occur. An example. In defining a future industrial area, the strategic plan document is not obliged neither to give the size of the building and the annexed areas, nor to give details of the allowed activities. Therefore, SEA cannot well evaluate anything or, at least, is able to make just an approximate evaluation from which, in most cases, no problematic profiles emerge. Not because they are absent, but because the object of the assessment is deprived of those crucial elements that characterize the possible environmental damage. In this case, Lombardy has acted more to protect its growth model than to care about the challenge that SEA had launched at the time.

4 Land Consumption in Lombardy: The Regional Law 31/2014

In 2014, the Lombardy publishes the Law n. 31, titled *Disposizioni per la riduzione del consumo di suolo e la riqualificazione del suolo degradato*.[2] It became the main regional law on the regulation of land&soil and, by consequence, of the land consumption. Considering the study done on the evaluation of the land consumption in SEAs, we are going to analyze two of the main interesting aspects of this law: (1) the definition of land consumption and (2) the indicators to calculate the land consumption.

Starting from the former, in the article n. 2 we can find the definition of *land consumption*:

> c) consumo di suolo: la trasformazione, per la prima volta, di una superficie agricola da parte di uno strumento di governo del territorio, non connessa con l'attività agro-silvo-pastorale, esclusa la realizzazione di parchi urbani territoriali e inclusa la realizzazione di infrastrutture sovra comunali.[3]

According to the law, the land consumption is the transformation, for the first time, of an agricultural surface, decided by each local urban plan. It is important to highlight that in this paper, and in the law too,[4] with the term 'agricultural surface' we consider the surfaces formed by agricultural and woodland areas.

Moving to the latter, continuing in the article n. 2, the law defines how to calculate the land consumption:

---

[2] *Disposition for the reduction of the land consumption and requalification of the degraded soil.*

In the next notes we translate the definitions provided by the law: the translation is not completely clear since also in Italian the definition is complex and a specific jargon is used.

[3] Land consumption: the transformation, for the first time, made by the urban municipal plan, of an agricultural surface not connected to woodland activities, not including new urban parks and including the realization of over local infrastructures.

[4] The agricultural surface is defined in the RL 31/2014, Art. 2, clause 1, letter a.

c) [...]il consumo di suolo è calcolato come rapporto percentuale tra le superfici dei nuovi ambiti di trasformazione che determinano riduzione delle superfici agricole del vigente strumento urbanistico e la superficie urbanizzata e urbanizzabile.[5]

The land consumption is calculated as the percentage ratio between the surface of the new areas of transformation, who determine reduction of agricultural surfaces, and the urbanized surface plus the 'urbanizable'[6] surface. The formula of this indicator, that we will call index of land consumption, ILC, is the following:

$$\text{ILC} = \frac{(\text{New areas of transformation})_{\text{agr}}}{\text{Urbanized surf.} + \text{Urbanizable surf.}}\%$$

The law gives also the definition of urbanized and urbanizable surface, as:

b) superficie urbanizzata e urbanizzabile: i terreni urbanizzati o in via di urbanizzazione calcolati sommando le parti del territorio su cui è già avvenuta la trasformazione edilizia, urbanistica o territoriale per funzioni antropiche e le parti interessate da previsioni pubbliche o private della stessa natura non ancora attuate.[7]

Urbanized and urbanizable surface consist of the areas already urbanized in the past or that are going to be urbanized.

This definition leaves space for some doubts and analyzing the formula of the ILC it is possible to better understand the issue.

Firstly, we can see the numerator: the new areas of transformation. The word new implies that just the urbanizations decided by the last version of the urban plan are taken into account. Actually, the natural surfaces that are going to be transformed in urban areas, generating a future land consumption, are also those ones decided by old urban plans. These old areas of transformation are still not realized and are confirmed by the new urban plan. Nevertheless, they are not regarded as areas, whose environmental impacts have to be evaluated by SEA. According to the law, they are treated as they were urbanized even if they are not. If you consider the real land cover, they appear as natural surfaces. Thus, they are still operating all of the ecosystem services.

Moreover, it is interesting to note that new areas of transformation is a precise term in this context, which identify only those areas located out of the so-called urbanized area. Not all the transformations decided by the urban plan, also considering just the new one, happen out of the urbanized areas. If anything, several of them are located into the urbanized areas. This would mean that these transformations are not treated

---

[5]The land consumption is calculated as the percentage ratio between the surface of the new areas of transformation, which determine reduction of agricultural surface, according to the urban plan in force, and the urbanized and 'urbanizable' surface.

[6]We will use this word, even if it does not exist in English since is the simplest translation of the Italian word. We will explain in depth what the law considers as 'urbanizable (that has to be urbanized) surface'.

[7]Urbanized and urbanizable surface: the areas already urbanized or on going to be urbanized, calculated by making the sum of those areas of the territory where the transformation for human functions was done and those areas involved in transformations, public or private, still not realized.

as land consumption, but rather just as regeneration or modification of urban areas. This appears true in theory, not in the reality. Indeed the urbanized area is a concept related to a definition provided by another regional law (RL 12/2005). There, it is defined the so-called 'tessuto urbano consolidato', TUC, that can be considered as the urbanized area politically defined or designed by each Municipality.[8] In theory, it should coincide with the real urban area. In practice, what the urban plan defines urban areas is not always what is really urban (Fig. 1).

The analysis performed validates the result obtained, showing that the Concrete What the law considers urbanized area is not really urbanized. This implies that also the inner transformations can generate land consumption. In the formula of the ILC they will be not counted. And SEA cannot catch them, consequently.

Secondly, looking at the denominator, we find the *urbanizable surface*. The law does not declare what it is exactly the *urbanizable surface*. According to other official

Fig. 1 The ambiguous overlapping between **a** urban surface, according to the land cover data set (Dusaf 201511), in red, and **b** the boundaries of the urbanized area, defined as TUC, the black line. Note that inside the TUC many areas are not sealed, they are called urbanized areas but they are not really urbanized. *Source* Elaboration from the TUC of the municipality of Sesto Calende

---

[8]The official definition of the TUC is provided by the RL 12/2005, art. 10, clause 1, letter a. According to the law, it is the document of the local set of planning rules that defines each municipal urbanized area.

documents,[9] it is possible to argue that it represents the transformations located into the *urbanized areas*, the TUC, and the transformations already decided by old urban plans, still not realized. Hence, these areas are natural areas (rural or woodland settings) and the new urban plans decided to confirm their future transformation in urban areas.

$$\text{Urbanizable surf.} = (\text{Inner tranformations} + \text{Transformations from the old urban plan})_{agr}$$

As discussed above, the point is that even the urbanizable surface in the reality generates evident land consumption.[10] The formula included into the law does not consider this aspect, assuming that just those transformations, new and outside the urbanized area, are regarded as land consumption. If we wanted to calculate really the land consumption, we would reformulate the ILC in this way:

$$\text{ILC} = \frac{(\text{New areas of transformation})_{agr} + (\text{Urbanizable surf.})_{agr}}{\text{Urbanized surf.}}\%$$

Finally, there is the urbanized surface. The 'trap' in this case is that the law does not give any official indications about the source of data useful to define and calculate the urbanized surface. On the one hand, we could consider the data set of the use and land cover where the cover of the areas are mapped.[11] In this way, we will take into account what is de facto urban. On the other hand, we could consider another data set which is elaborated by each Municipality who identify its urbanized area, the TUC. In this way, we will take into account what is *de iure* urban. And as explained before, these two data do not overlap. Changing the data set, which is not officially indicated in the law, we will obtain different results.

In addition to this index, in the law it is presented another indicator to measure the land consumption. It is the index of reduction or variation of land consumption, called also ecological balance (EB), and it is defined as:

> d) la differenza tra la superficie agricola che viene trasformata per la prima volta dagli strumenti di governo del territorio e la superficie urbanizzata e urbanizzabile che viene contestualmente ridestinata nel medesimo strumento urbanistico a superficie agricola. Se il bilancio ecologico del suolo è pari a zero, il consumo di suolo è pari a zero.[12]

---

[9]We are referring to the 'Piano Territoriale Regionale d'Area (PTRA) della Franciacorta', 2017 and to 'Relazione annuale sullo stato della pianificazione territoriale in Lombardia 2016', Eupolis 2016.

[10]It is interesting to highlight that, already in 2013, the urbanizable surface was the 58% of the total amount of the transformations in the Lombardy Region (Pileri 2005: p. 109). This means that the most part of the transformations were not evaluated in SEAs.

[11]In the Lombardy Region the data set on the land cover is the DUSAF, available for the years 1999, 2005, 2007, 2009, 2012 and 2015. The equivalent scale is 1:10.000.

[12]The difference between the agricultural surface that is going to be transformed for the first time by the urban plan, and the urbanized and urbanizable surface that, at the same time, is transformed in agricultural surface by the same urban plan. If the ecological balance is zero, then the land consumption is zero.

The EB index is obtained by making the difference between the agricultural surface that will be transformed in urban area, according to the decisions of the new urban plan, and the urbanized and urbanizable surface that will be transformed in agricultural area, according to the decisions of the same urban plan. The formula will be:

$$\text{BE} = \text{Agricultural surf.transformed} - \text{Urbanized and urbanizable surf.transformed}$$

By this definition, if the ecological balance, BE, is zero, then, the land consumption is zero.

It is possible to note some critical elements of this definition. The first lies in the concept itself of balance. Even if we assume that some urban surfaces are reconverted in agricultural or natural areas, the real new land consumption of agricultural surfaces will remain. Moreover, from the ecological point of view, the reconversion of urban areas in natural ones do not generate the same values of an 'old' natural areas (e.g. as happened for the woodlands). Another critical element that appears from the study of several SEA and urban municipal plans, is what we can consider as urbanized *and urbanizable surface transformed in agricultural areas*. Indeed, considering the definition explained before of urbanized and, overall, urbanizable surface, some plans insert in the calculation of the ecological balance some old transformation areas, still not realized, that they decide to delete. In other words, in this case, we do not have any renaturalization of urban areas, but rather, we have agricultural surfaces, that had to be transformed in urban areas according to the decisions of the old urban plan.

5 The Analysis: 100 SEAs for 100 Municipal Urban Plans

In order to understand whether the regional law 31 has positively affected the urban planning in Lombardy regarding the evaluation on the issue land&soil, an analysis of 100 SEA was carried out. The selection of SEA responds to the necessity of considering only documents elaborated after the emanation of the law. Therefore, the sample of the analyzed SEA has been made as following:

- 33 SEA approved in 2017;
- 46 SEA approved in 2016;
- 21 SEA approved in 2015.

Looking at the typology of municipality, we considered different urban sizes looking at number of inhabitants (Fig. 2):

- 40 SEA related to municipalities with less than 2.000 inhabitants;
- 22 SEA related to municipalities with more than 2.000 and less than 5.000 inhabitants;
- 20 SEA related to municipalities with more than 5.000 and less than 10.000 inhabitants;

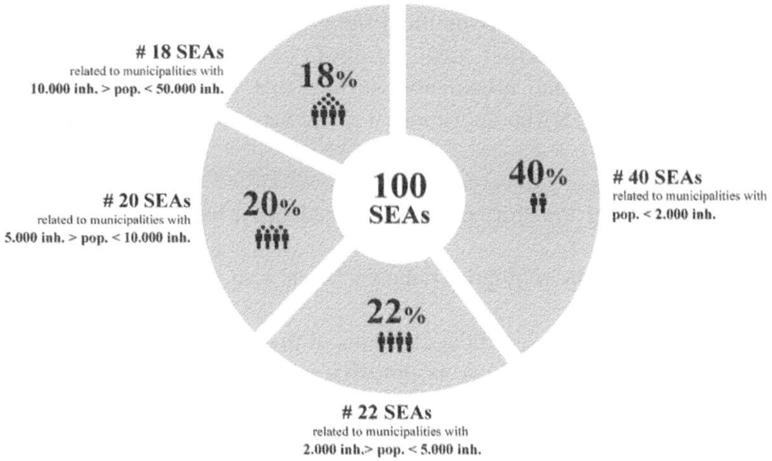

Fig. 2 The sample of SEAs analyzed and the demographic size of the municipalities related

- 18 SEA related to municipalities with more than 10.000 and less than 50.000 inhabitants.

The analysis was conducted by studying each SEA. In particular, it was considered how each SEA has evaluated the land consumption generated by the local urban plan. The main steps faced in the analysis were:

- Step 1 → the presence of an evaluation of the land consumption through indicators;
- If Step 1 is 'yes', then Step 2 → analyze which kind of indicator was used to evaluate the land consumption.

Starting from the first step, 43 SEA (43%) do not give any comment or evaluation through indicators about the land consumption generated by the decision of the urban plan. The others, 57, (57%), tackle the issue of the land consumption and evaluate it through indicators.

Moving to the second step, on the 57 SEAs, 44 of them (77%) display one indicator, and 13, the 23%, provide more indicators: 10 of them propose two indicators, while 3 propose three indicators. These 73 indicators ($44 + 10 \times 2 + 3 \times 3$) do not use only the ILC or the EB indicators provided by the law. There are six different typologies of indicators, as following:

1. index of land consumption proposed by the Regional Law 31/2014—ILC1 (9 cases, 12%)
2. index of reduction or variation of land consumption, called also ecological balance, proposed by the Regional Law 31/2014—EB (6 cases, 8%)
3. index of land consumption—ILC2 (32 cases, 44%)
4. index of agricultural land consumption—ALC (3 cases, 4%)
5. index of sprawl (4 cases, 6%)
6. other indices (usually not clearly explained in SEA) (19 cases, 26%).

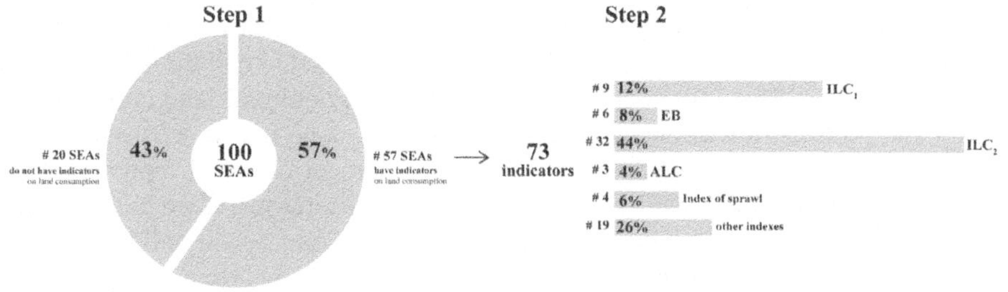

Fig. 3 The distribution of the different indicators used by the sample of SEAs

The first four indicators are considered in a quantitative approach, that means that in SEA it is specified how they were calculated. Others, the last two typologies, are presented in SEA with a qualitative approach, without any explanation or data related.

According to these six different typologies and remembering that some SEA propose more indicators, we will obtain what is described in Fig. 3.

In order to understand the diversity between these indicators, it is useful to describe each of them. The first two indicators, the ILC_1 and the EB are proposed by the regional law, and they were already described in the previous paragraph. The third one is a sort of customization of ILC_1 that we will call ILC_2: it is the most common. It represents the percentage ratio between the urbanized surface and the total surface of the municipality. It corresponds to the urbanization coefficient, belonging to the most known family of composition indicator, and the formula is:

$$ILC_2 = \frac{\text{Urbanized surface}}{\text{Total municipal surface}}\%$$

The urbanized surface considered in this formula is different from the one used in the denominator of the ILC_1. In this case the urbanized surface is the sum of the present and the future one, after the transformations decided by the urban plan.

The fourth one is the index of agricultural land consumption. The aim of the index is to evidence how much the new urban transformations will affect the agricultural land. In SEA there were found different ways to calculate this index, that are:

$$ALC_1 = \frac{\text{Agricultural surface}}{\text{Inhabitans}}$$

$$ALC_2 = \frac{\text{Agricultural surface}}{\text{Total municipal surface}}\%$$

$$ALC_3 = \frac{\text{New transformation areas}}{\text{Agricultural surface}}\%$$

These formulas show two of the main relevant issues related with the indicators of land consumption. On the one hand, the same name is used to call three completely different indexes. Thus, if we compare the index of agricultural land consumption ALC_2 and ALC_3, we will obtain a false relation: the same index, according to the name, consider different elements and the value percent obtained represents two different values. On the other hand, these indexes are useful just if we consider two different times in order to compare the present situation and how it will be changed by the new urban transformations. For example, for the ALC_2 we need to consider the value of the index at the present time, ALC_2 (t_0), and in the future, ALC_2 (t_1), assuming that the urban transformations decided by the plan will be realized. In particular the index ALC_2 corresponds with the agricultural coefficient and it is not directly related with the measure of land consumption.

The index of sprawl is not clearly explained in SEAs analyzed. We do not have any formula in order to understand how it is exactly calculated. SEAs explain it in two different ways. On the one hand, as the tendency of the urban dispersion, considering the percentage of the new transformations areas, which is directly in touch with the urbanized surface. On the other hand, the index of sprawl makes a distinction between the transformation areas, that are included in the urbanized surface already existing and the others, that generate a risk for the agricultural surfaces. The aim of both indicators is to avoid the dispersion of new urban areas, proposing a concentration of the new transformation areas or a saturation of the empty spaces into the already urbanized zone.

Finally, there are SEAs that propose indicators, just citing an index, without any explanation or formula, useful to understand what they are consider. The name of the indicators always changes and there are no evidences to identify the precise index they are considering. Moreover, there is not a quantitative approach and, hence, SEAs just cite the issue of the land consumption.

6 Rethinking the Role of SEA

SEA is a planning tool, useful to control the actions on the territory proposed by the urban plans at different scales, macro and local ones. As well explained by Sabeva (2015: p. 101):

> SEA is one of the basic tools for preventive control and is also an integral part of the process of development and adoption of plans, programs and politics. Its main objective is achieving a sustainable development, minimizing the adverse impacts on environment and improving its quality. SEA should ensure that plans and programs take into account the potential environmental impacts they cause.

Starting from this perspective, we elaborated an analysis in order to understand whether SEA, in the Italian context, is effective on the issue of the containment of the land consumption. The premises related with the specific conditions of the case study are:

- the context is the Lombardy, in Italy, where the main transformations are managed at the local scale, by the municipal urban plan. Thus, the 100 SEAs considered are those ones that have to evaluate the municipal plans;
- there are recent regional laws on the urban planning, LR 12/2005, and on the land consumption, LR 31/2014. They should clarify some key elements, useful to make uniform the local planning of more than 1500 Municipalities. By analyzing in depth these laws, several ambiguous definitions emerge (consider the third and the fourth paragraphs);
- the RL 31/2014 provides two indicators to measure the land consumption, the index of land consumption, ILC, and the ecological balance, EB;
- the sample is formed by 100 SEAs elaborated after the regional law 31/2014.

Assuming these starting points, it was investigated how SEAs evaluate the land consumption and, when they do, if they use the method introduced by the law. The result of the analysis is that, on a total of 100 cases, just 15 SEAs, the 15%, adopted the indexes of the law. In the other cases, several alternative indexes are used or, in the worst situations, the land consumption is not evaluated at all (44 cases). This data demonstrates that each municipal urban plan, and each relative SEA, calculates with own criteria the land consumption. Such irregularity in the choice of the calculation's method implies several weaknesses, which we can be summarized in:

a. the difficulty in reading the plans. If each plan is different, we are obliged to understand the specific case and method of calculation and it is not obvious that the plan clearly explains the methodology used. In other words, if the plan or SEA does not provide sufficient explanations, we are not able to understand the calculations and the data;
b. the impossibility of checking the data indicated in the plans and in SEA. It is usually really complicated to replay the calculations;
c. the lack of comparison between different plans and SEA.

These weaknesses contribute in decreasing the effectiveness of SEA as real evaluating and planning tool. And the recent law was not able to enhance the clarity of the methods. On the one hand, it does not force to adopt its method to calculate the land consumption. On the other hand, even the law is not completely clear in its methods too, as explained above.

From the analysis, it emerges how one of the main critic points is the ambiguity of the language and calculations methods both of SEAs and of the laws. Such *ambiguity* seems to be a characteristic of the urban planning (Pileri 2018): it contributes in generating confusion and in leaving space to imprecisions and approximation. In this context, the goodness of SEA (and of the urban plan too) is completely referred to the politician or to the planner involved in the decision's processes. This is not enough considering the urgency of reducing the land consumption. Therefore, providing new, uniform and clear, definitions and methods, or rethinking better the old ones, could be the first action, necessary to make SEA an effective tool of evaluation, control and support of the planning process.

Acknowledgments The analysis presented in this paper was elaborated taking into account the master thesis by Micol Bellinetti, 'La valutazione ambientale strategica e il consumo di suolo: analisi dell'efficacia delle VAS in 100 comuni lombardi', 2018, Politecnico di Milano, supervisor Paolo Pileri, assistant supervisor Rossella Moscarelli. In this document t is possible to see 100 SEAs in detail.

The authors are both responsible of the whole paper that they have thought and edited together, even if the paragraphs 1, 2 and 3 can be conferred to Paolo Pileri and the others to Rossella Moscarelli. the conclusions to both authors.

References

AA.VV (2015) The Soil Atlas 2015. Facts and figures about earth, land and fields, Heinrich Böll Foundation, Berlin and Institute for Advanced Sustainability Studies, Potsdam, Germany

Munafò M (2018) Consumo di suolo, dinamiche territoriali e servizi ecosistemici, Rapporto 288/2018. ISPRA, Roma

Pileri P (2005) Che cosa c'è sotto. Il suolo i suoi segreti le ragioni per difenderlo, Altreconomia, Milano

Pileri P (2018) 100 parole per salvare il suolo. Piccolo dizionario urbanistico-italiano, Altreconomia, Milano

Sabeva M (2015) Strategic Environmental Assessment—necessity, principles and specificities. In: Review of applied socio—economic research, vol 9, no 1

Sini V (2016) Repertorio della normativa in materia di VAS. Normativa delle regioni e province autonome, ISPRA, Roma

Treville A (2011) Strategic Environmental Assessment as a tool for limiting land consumption. In: Special conference on Strategic Environmental Assessment, IAIA SEA Prague II, 21–23 September 2011

A Tourist Territory Between the Lagoon and the Sea: A Strategy for a Dynamic and Lasting Development

Otello Bergamo

Abstract The essay deals with the analysis of the city of Jesolo, as a territorial area with a high tourist vocation characterized by particular geographical areas with high environmental and landscape quality. A sustainable tourism development strategy is identified based on six actions such as the defense and upgrading of the coastline, the regeneration of first-rate coastal settlements, the protection and safeguarding of natural environments and contexts of historical value, the re-naturalization of portions of areas of more recent reclamation of limited agricultural value and the redevelopment of residential areas.

Keywords Tourist territory · Lagoon · Sustainable development · Statistics

1 Introduction

In 2017 there were about 113,500 accidents in urban areas in Italy, some of which were caused by poor road surface conditions due to inefficient maintenance and safety of the road network (Aci–Istat 2017). The problem is especially acute in urban areas, where deteriorated infrastructures, obsolescent facilities and serious congestion problems are resulting in economic loss, environmental damage and societal harm (Aci–Istat 2017).

The city of Jesolo can be considered in all aspects the tourist capital of the Adriatic, an important reference pole known above all as a city of entertainment and less under the most important and peculiar aspect of the place, the environment. During the 1900s, first the complete reclamation and then the seaside tourism gave the territory of Jesolo a vast agricultural area and a large tourist promenade, radically modifying the economy, the environment and the landscape.

Along the 12 km of coastline, three specific homogeneous territorial areas can be identified:

O. Bergamo (✉)
St. Eng, Venice, Italy
e-mail: st.eng@libero.it

G. Campeol (ed.), *Strategic Environmental Assessment and Urban Planning*, Green Energy and Technology, https://doi.org/10.1007/978-3-030-46180-5_8

- the first to the west includes the areas belonging to the northern lagoon of Venice and the territories located on the right Sile (Piave Vecchia). The lagoon area is composed of a series of fishing valleys and rivers of great environmental and ecological value.
- the second corresponds to the area of the river Piave backbone, which houses the urban centers of the hinterland and is the last part of the great ecological corridor that connects the Alpine territories to the sea.
- finally, the central area, which involves a much larger space and includes the urban center of Jesolo and the tourist town of Lido di Jesolo, connected by a significant infrastructure system and large agricultural areas.

Important changes have been made on each of the three areas, and important transformations are underway: agriculture, seaside tourism and demand for free time are changing, the system of territorial mobility is being strengthened and that of accessibility to the coast is being integrated. It is fundamental to understand the role of Jesolo in this process of change, to positively orient the pushes to an environmental renewal (Fig. 1).

In the city of Jesolo all the themes of the territorial planning, on which planning is being measured in the Veneto region, have a specific and non-homologous declination to that of other realities, even nearby. The strong relations of the infrastructure sytem of Jesolo (Fig. 2) with the central metropolitan system of Veneto, which includes the foothills and the cities of Padua, Treviso, Vicenza and Venice, highlight their

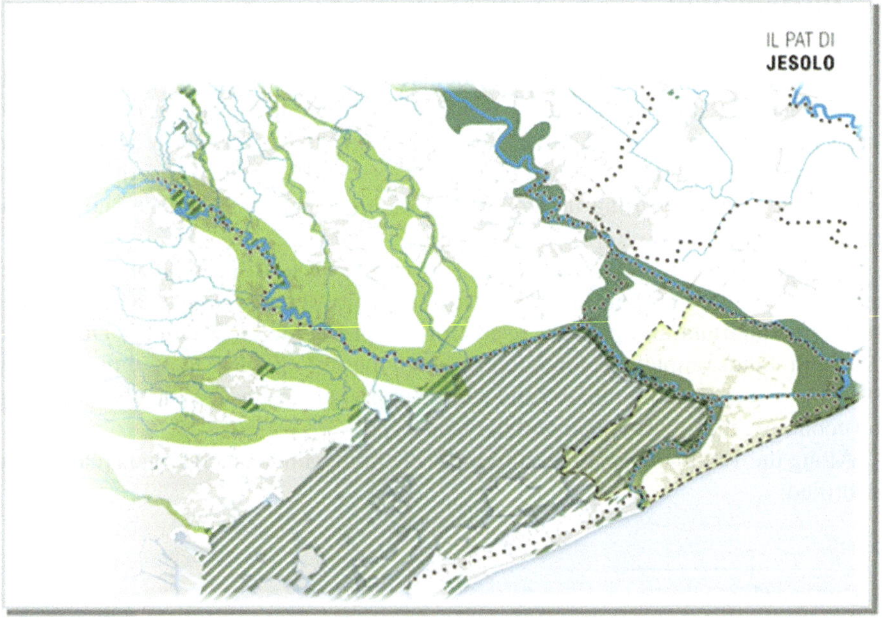

Fig. 1 The environmental reference system of the Jesolo territory

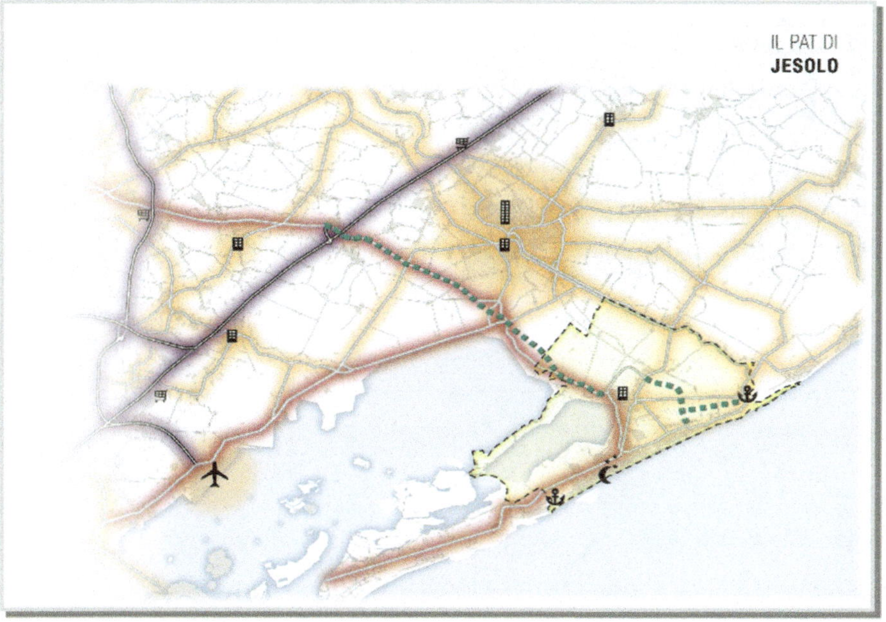

Fig. 2 The infrastructure system of reference of the Jesolo territory

peculiarities, configuring it as the bathing city of the central metropolitan area of the Veneto, one Natural "outlet" to the sea by the central areas of Veneto.

2 The Specificity of the Lagoon Area

The lagoon area is formed by an alternation of water bodies and embankments and that shape draw the template of the lagoon. In this context, the only elements of a punctual settlement are the buildings located along the Sile and those are historical buildings situated near the fishing area.

If the system of the valleys and of the Lagoon, though generated by an intense and systematic action of anthropization, has remained for a long time unrelated to the dynamics of perception and enjoyment of the seaside resort, it is evident that the integration of the tourist offer can count on a vast and evocative environment that can contribute both to the enrichment of the bathing product and to the deseasonalization of the flows, respecting the habitats and the vegetal, fish and fauna resources.

The fishing valleys are a true cultural landscape, result of the anthropic transformation. A territory that carries within itself original cultural traditions and attitudes, linked to the particular exploitation of the wetlands.

The challenge is to combine the dynamics of evolution of these environments of high natural value with the energy and resources necessary to ensure maintenance

Fig. 3 The defense and enhancement of the lagoon area is the strategic objective of the PAT

and care. The operation of territorial regeneration can lead to the formation of a unique image in the coastal context, usable especially during the low season (spring and autumn), contributing to extend the seasonality of the territory (Fig. 3).

3 The Sustainable Requalification Strategy

In this context of progressive change, it is a matter of identifying the fundamental values to be protected and promoted, supporting agriculture, regenerating the population centers, transforming tourism from a coastal and coastal event to a territorial phenomenon, using the waterways, the Rivers Sile and Piave, the Litoranea Veneta as backbones of change, reducing pressure on the coast and expanding the tourist offer and seasonality.

- Therefore, actions are taken of:
- Defense and redevelopment of the coastline;
- Regeneration of coastal settlements of first plant;
- Protection and protection of natural environments and historical contexts;
- Environmental renovation of portions of areas of more recent reclamation of limited agricultural value;
- Upgrading of population centers.

It is a question of giving the coastal tourism system an appropriate infrastructural, environmental and service background, as happens in similar coastal areas, where seaside resorts and settlement system are intertwined with the territorial scale (Holland, France, and in the United States the Florida states and California) restoring ecological connections, redeveloping degraded landscapes, strengthening the functional structures for pleasure boating for navigation both in maritime waters and in internal waters.

These are not isolated actions, but a line of territorial redevelopment that unites the entire tourist district of the Metropolitan City of Venice, which the PAT di Jesolo declines according to the local peculiarities.

In fact, the transformation of tourist facilities and equipment, the evolution of agriculture, the reorganization of the system of accessibility to the sea, covers the entire territory of the Venetian east coast, a flat hinterland, which extends for about forty kilometers between the northern lagoon of Venice and that of Marano, engraved by Alpine rivers, furrowed by road axes of relationship between the arc of the urban centers of the foothills, that of the centers of the second band, and that of the local coastal centers.

The morphological and logistical margin of this district is determined by "Mediterranean Corridor", located a little more than twenty kilometers from the sea and strengthened by Marco Polo International Airport.

In this territory, already now, tourist function and residential function are not intertwined only along the coast, but directly involve the second-tier centers, up to the infrastructural arch of the hinterland. This segment of the Veneto logistics platform, the highway, railway and airport infrastructures, currently being completed, should not be considered only with a way to bring the Po Valley and Europe closer to the holiday city, to bring vacationers to the sea in the Po Valley and Central Europe, but on the contrary as a way to unite the Eastern Venice with the central Venetian underground platform and Europe.

On the other hand, if we consider the economy of this territory, observing it from the point of view of consumption, residents and tourists, we can add new considerations to those developed so far.

We can estimate the annual expenditure in Jesolo in about € 1.1 billion, of which 34% due to the spending of the residents (valued at about € 40 a day) and 66% to the spending of tourists (valued at about € 95 the day). Added to that of Cavallino-Treporti, the total annual expenditure reaches 2 billion euro, and is equivalent to the one obtained for Venice, adding to the expenditure equivalent to 10 millions of this tourism, that of about 56,000 residents in the Historic Center (Fig. 4).

Naturally these values indicate a structural problem: in the eastern coast about 70% of consumption is concentrated in a short period of the year, generating strong imbalances on the social and economic system. Completely opposite, is the result in Chioggia-Sottomarina, where the share of tourist consumption would represent about a quarter of the total (Fig. 5).

The east coast of the Metropolitan City of Venice should be considered unitarily as a privileged place, where to live, from which to work or holiday. Not only a

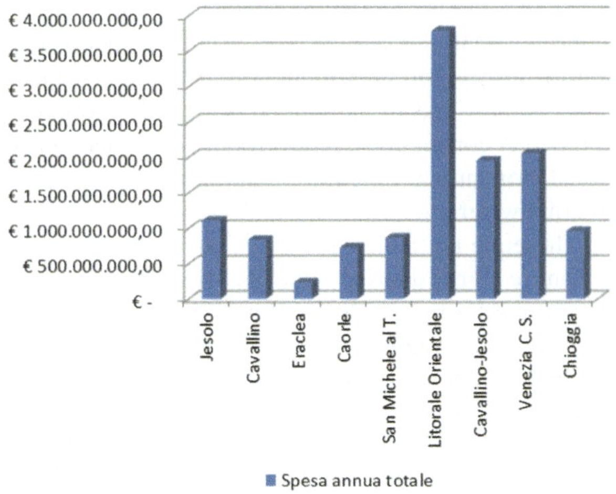

Fig. 4 Total annual expenditure

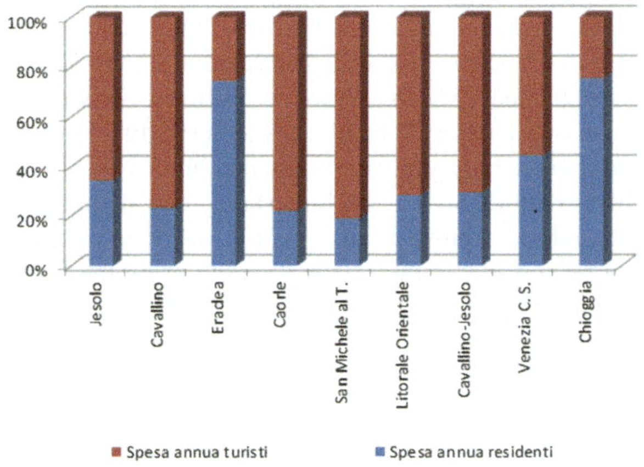

Fig. 5 Percentage of annual spending by tourists and residents

cosmopolitan district but also an open residential park of great environmental quality, which extends from the linear coastal city to the second-class historical centers, alternating natural habitats, residential areas, tourist settlements, infrastructural corridors, wide naturalization areas, areas of excellent productions. With the Litoranea Veneta organized as a hinge and formal prospect of the first tourist line to the hinterland: navigable park along a system of canals that allows you to travel on the inland waters (parallel to the coast) from Monfalcone to the Po; relationship plug that links channels, landings, docks, recreational infrastructure and tourism services.

4 Conclusions and Future Research Perspectives

In order to have greater synergy between development and protection of natural resources, it is necessary to plan environmental protection programs over the medium to long term.

In the current context of development, linked to the concept of sustainability, a central role is determined by the public function at all levels, regarding territorial planning, public-private coordination and monitoring of the country's industrial, economic and environmental development.

The definition of a sustainable city passes through the drafting of an improvement of the environmental impacts on the territory in the perspective of sustainable tourism connected to the development of a slow tourism through the enhancement of the lagoon allowing an osmosis and a dialogue between the built part and the naturalistic one of the city.

Jesolo, being a tourist-bathing resort, has always valued above all the seaside area to encourage tourism and development, but planning a revaluation of the eastern part of the Venetian Lagoon will be able to appreciate a natural oasis otherwise hidden for the purpose of a controlled, modern and innovative tourist enjoyment that allows the city to have a balanced impact between summer—winter access and permanence, making the city known from a different perspective from the classic sun and sea.

In this context, the valorisation of the blue flag for the beaches, conferred for the fifteenth consecutive year, associated with the development of techniques and procedures controlled from an environmental point of view, starting with the differential collection, will allow the city a wider scope in the tourist and residential offer of the locations.

References

Aci–Istat (2017) Incidenti stradali. Roma

Badiani B, Badiani G, Badiani L (2011) Jesolo 2012. The City Beach, Archivio di studi urbani e regionali

Bergamo O, Ceccato F, Pezzato D (2017) Subsidenza nel litorale veneto: analisi del fenomeno nel comune di Jesolo. Jesolo, Comune

Grigoletto M, Lisi F (2013) Complex models and computational methods in statistics. S. Petrone. Springer. ISBN: 8847028701

Jesolo (2002) La città del tempo libero: Jesolo fra presente e futuro. Opus Ed

Methodological Integrations Between SEA and Plan "Design". The Case of the Urban and Regional Planning for the Municipality of Jesolo

Giovanni Campeol, Cristina Benvegnù, Erblin Berisha, Lorella Biasio, Sandra Carollo, Silvia Foffano, and Davide Scarpa

Abstract The paper reflects on the capability of the SEA to integrate various evaluation levels elaborated for the urban and environmental planning. The document deals with the case study of the SEA laid down for the Land Use Plan (PAT) of the Municipality of Jesolo. The environmental evaluation process has permitted to elaborate two specialized evaluation studies in the field of landscape planning and of the recovery and use of Jesolo Lagoon.

Keywords Integration environmental assessments · Jesolo · Urban and environmental Planning · Landscape evaluation

G. Campeol (✉)
Università Iuav di Venezia, Venezia, Italy
e-mail: giovanni.campeol@gmail.com

C. Benvegnù · L. Biasio · S. Foffano
Alia Office, Treviso, Italy
e-mail: cribenvegnu@gmail.com

L. Biasio
e-mail: lobiasio@alice.it

S. Foffano
e-mail: silvia.foffano@hotmail.it

E. Berisha
Department of Regional and Urban Studies and Planning DIST, Politecnico di Torino, Turin, Italy
e-mail: erblin.berisha@polito.it

S. Carollo
Università degli Studi di Padova, Padova, Italy
e-mail: dott.sandracarollo@studiocaldato.it

D. Scarpa
Natural Science, Mira (VE), Italy
e-mail: davidescarpa@gmail.com

G. Campeol (ed.), *Strategic Environmental Assessment and Urban Planning*, Green Energy and Technology, https://doi.org/10.1007/978-3-030-46180-5_9

1 The SEA as an Occasion for Integrating the Environmental Assessments

In 2017 The European directive no. 2001/42/CE,[1] improperly defined as SEA Directive—Strategic Environmental Assessment—, is a procedure/process for evaluating the sustainability of territorial programs and urban plans, through which the territory transformations are assessed beforehand in order to verify their level of interference (the impact, which can be positive or negative), with the three big systems that compose the environment (abiotic, biotic and human systems).

The adoption of this directive de facto put an end to a long legislative period in which the European Union was committed to define the procedures, methodologies and techniques for evaluating beforehand the environmental impacts that might be caused by projects, programs and plans. In fact, Directive 85/337/EEC on the Environmental Impact Assessment (EIA) of 1985 was followed in 1992 by Directive 92/43/EEC on the Assessment of Implications for European Sites (AIES), aimed at protecting the biodiversity of Sites of Community Importance (SCI).

The presence of the three evaluating procedures (EIA, AIES and SEA in issuing order), together with the landscape quality evaluation procedure in Italy, needed and still need procedural and methodological considerations with a view to integrate different evaluating levels for improving the ability to interpret complex environmental dynamics. From a methodological viewpoint it is therefore necessary to organize procedures and develop methodologies able to integrate different evaluation levels, which can be best placed within the Strategic Environmental Assessment system.

2 The Case of SEA in the Municipality of Jesolo: An Example of Methodological Integration

In the experiences realized in almost 15 years of the application of the EU directive and the transposition into the Italian national law, the SEA procedure has not always improved the planning instrument with the introduction of environmental elements (abiotic, biotic, human). The analysis of many Environmental Reports (basic documents for the SEA procedure) on regional and national strategic Plans[2] has highlighted a general analytic redundancy produced by an excessive collection of environmental data that, on the other hand, cannot always be connected with the urban tool. Moreover, a very redundant analysis is followed by poor evaluations, essentially slightly connected with the planning strategies and often prevalently used to justify the choices made in the plan.

[1]Directive 2001/42/EEC of 27 June 2001, published in the O.J.E.C. n° L 197 of 21 July 2001.

[2]*National Operational Programme and Action System ESF* 2007-2013 Axis E: *Institutional Capacity*—Specific Objective 5.5: *Strengthening and integrating the environmental governance system* Action 7B: *Actions supporting SEA processes and EIA procedures.*

The Environmental Report elaborated for the SEA in the land-use plan of the municipality of Jesolo tries to go beyond the limits highlighted in these experiences, first of all by selecting and sintering the components and environmental indicators from the big quantity existing in literature and referred to by the Veneto Region, on the basis of their possible relations with the strategic territorial planning. Nevertheless, the same report is particularly massive, also due to the demands for in-depth analyses (not always welcomed) of the different environmental authorities involved in the SEA procedure.

Moreover, this Environmental Report is confirmed by three very in-depth analytical and evaluation studies that have made it possible to implement the urban and regional plans considering the environmental and landscape aspects, and also to integrate the evaluations into a single methodology.

In fact, Jesolo presents special geographical and natural conditions for the presence of varied landscape and natural characteristics, which make the area an emblematic case. In this sense, evaluating the different aspects within a single framework is an important basis for a planning of high environmental quality. In 2014 the complexity and variety of the territory in the municipality of Jesolo led to a More detailed landscape assessment of the urban and regional planning and, in August 2015, to a Leading Scheme of the Plan for tourism in Jesolo Lagoon. Furthermore, the Study for Implications on the European Sites was laid down with reference to the natural aspects.

2.1 In-depth Analysis of the Landscape Evaluation of the Urban and Regional Planning (PAT)[3]

The paper of the in-depth analysis of the landscape evaluation of the urban and regional planning (PAT) is part of the convention for the implementation of the PAT Overall Framework for the landscape aspects (Legislative Decree of the Veneto Region No. 42/2004), the promotion of the "archaeological landscape" in Venice Lagoon and the restoration of the seriously affected and degraded areas.

The objective is to analyze the territory of Jesolo municipality in order to restore the landscape, together with the archaeological sites, and consequently to enhance important areas of historical, cultural, landscape and archaeological interest. To this end, Annex A to the Convention (DGR No. 2619 of 30 December 2013) has been referred to, since it highlights how "the effective or potential integration of the archaeological assets into the environmental, historical, cultural heritage and the relevant services and their protection and rehabilitation may engender a specific added value that can be measured both in cultural and economic terms".

---

[3]In-depth analysis of the landscape evaluation of the urban and regional planning (PAT). Memorandum of understanding. Implementation of the PAT Overall Framework for the landscape art. 143, paragraph 1, Legislative Decree 42/2004, L. 16 April 1973, No. 171.

In the case of Jesolo this is an experiment, already methodologically applied in the case of the Urban Regional Planning of Evora (Portugal),[4] which has led to a definition of "the landscape planning objectives" spatially represented as "geographical scopes", rather than "urban areas or zones". This study also analyzes the issue of the "archaeological landscape" of Venice Lagoon by elaborating a specific table that highlights the objectives set in the UNESCO Management Plan "Venice and its lagoon"—2012–2018—and those of the Management Plan of the SCI/SPA sites existing in the lagoon.

In order to define the current situation of the landscape in Jesolo territory, also in relation to the different constraints, thirteen Landscape Scopes have been identified and outlined on the basis of two different types of readings, typological-legislative and qualitative. This methodology has permitted to identify homogeneous Landscape Unities (or Scopes), compared with the structural features from an urban viewpoint, within which one or more optical cones have been identified and then evaluated following the criteria set in Prime Ministerial Decree of 12 December 2005. The identification of these scopes has been drawn on the work of the Regional Planning Office of the Municipality of Jesolo, which has analyzed some areas—defined as badly affected—through a photographic survey. These places have been mapped and analyzed just like the other optical cones. The definition of the landscape characteristics of the different scopes has been carried out through an evaluation technique based on the perceptional reading of the landscape features, which has led to the definition of the landscape planning objectives, helpful to define the PAT strategies of Jesolo, also through the NTAs (Implementing Technical Standards).

However, it is necessary to underline that the discipline of urban planning has always considered the landscape plans through a two-dimensional map reading, starting from the wrong assumption that the landscape may be translated into space subdivisions (as in the case of traditional zoning). This approach has strongly reduced the two-dimensional translation of a three-dimensional geographic condition arising from the perception of the local culture.

The landscape can be evaluated following two interpretations, structural and perceptive. The former represents, through the two-dimensional analysis of the map formal characters (natural and anthropic networks, also through landscape ecology), the elements composing the form of the landscape, by translating them into landscape features, although with the above outlined limits. The latter represents, through the three-dimensional analysis, the view of the areas through images reproduced by identified optical cones in order to characterize the different geographic scopes existing in the municipal territory. The perceptive interpretation of the landscape cannot be translated into zoning, but rather into geographic scopes (Fig. 1), and it is particularly fit for evaluating an urban tool that produces strategic guidelines without translating them into a rigid map, such as the Urban and Regional Planning (PAT).

[4]Degree thesis "A model of landscape evaluation for urban planning: the case of Evora", supervisor Giovanni Campeol; advisors Leonel Fadigas, Sandra Carollo; graduate student Francesca Castagna; University IUAV of Venice, Faculty of Architecture, Degree Course in Architecture, academic year 2002/2003.

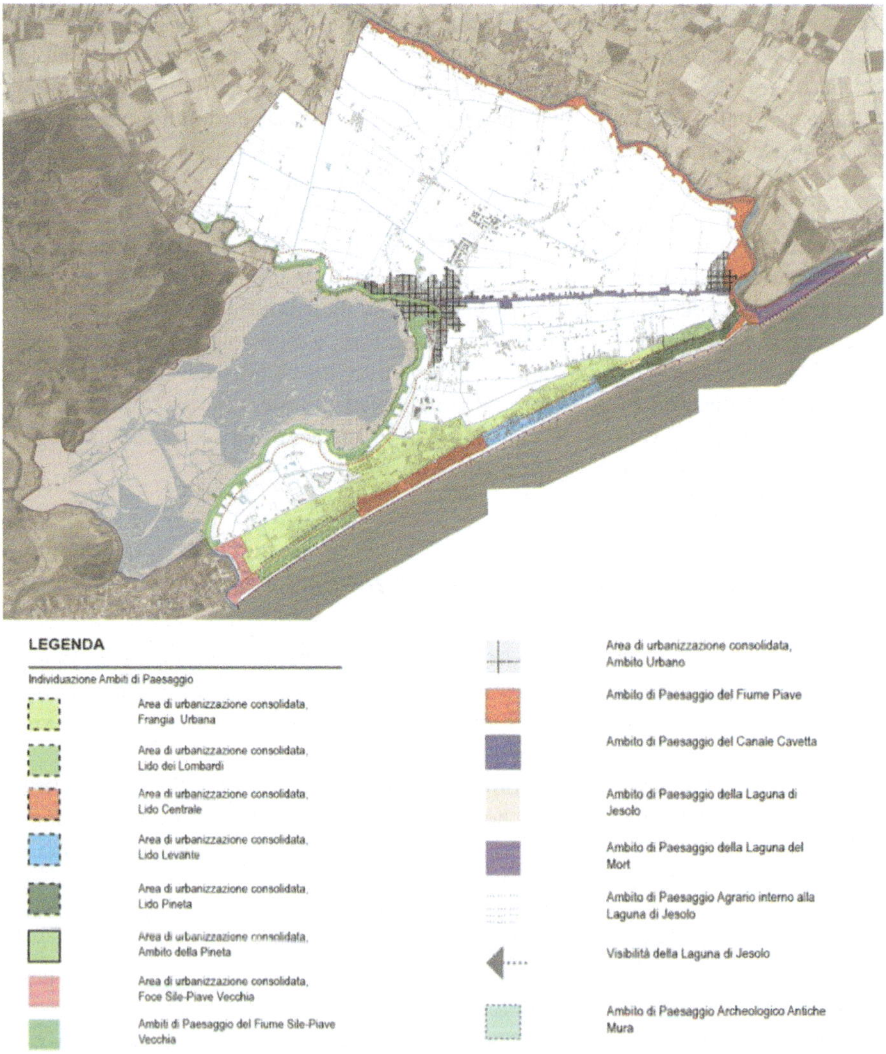

Fig. 1 Landscape scopes in the territory of Jesolo

The study also analyzes the "archaeological landscape" of Venice Lagoon by means of a table that highlights the objectives set in the UNESCO Management Plan "Venice and its lagoon" (2012–2018) and those of the Management Plan of the SCI/SPA sites existing in the lagoon. In fact, the landscape aspects and those related to the conservation of natural habitats play an important role in identifying the features of part of the lagoon in the municipality of Jesolo (Province of Venice).

In particular, the objectives of the UNESCO Management Plan that can be divided into 4 Action Plans (1—Protection and conservation of the heritage; 2—Sustainable use of the Site; 3—Communication, promotion and training; 4—Knowledge and sharing) also include objectives of the restoration and promotion of the archaeological heritage. The analysis of these issues has also led to the detailed elaboration of the paper called "Master Plan for Jesolo Lagoon as an in-depth analysis of the PAT", whose goal is to define a redevelopment project in this geographic scope that may enable the tourist to use the lagoon in synergy with the existing activities and elements.

In terms of archaeology, the optical cones have been set up for the archaeological site Ancient Walls of Jesolo, the only valuable site in the municipal territory. These overviews have been evaluated in order to define their landscape objectives, which have been in turn translated into projects. Therefore, a map concept and some views have been elaborated in order to define a new value to the "landscape characters" in the territory. This detailed evaluation has also enabled to reflect on the "effectiveness" and "legitimacy" of very wide constraints, i.e. the landscape and archaeological constraints imposed on Venice Lagoon and the coastal strip.

2.2 *The Master Plan for Tourism in Jesolo Lagoon*[5]

The municipality of Jesolo is the easternmost part of Venice Lagoon. This scope is totally private and has always been almost inaccessible to the public. This condition has prevented the local communities and especially the tourists from getting to know this extraordinary and somehow unique environmental resource.

The municipal authorities of Jesolo need to investigate in detail the issue of Jesolo Lagoon through the elaboration of the Master Plan on Tourism in Jesolo Lagoon, in order to strengthen the PAT strategic guidelines. This Master Plan is an orientation tool that defines the forms for tourist use of Jesolo Lagoon, all year long, in synergy with current activities as fishing, extensive fish rearing, hunting, etc., and coherently with the territorial and urban planning and programming tools (especially the UNESCO Management Plan of Venice and its Lagoon, and the Management Plan of the SCI/SPA Venice Lagoon), which make it possible to identify precise objectives for a development of sustainable tourism for Venice Lagoon and consequently for Jesolo lagoon.

The actions established in the document take into account the peculiar environmental features of the lagoon, especially hydraulic, geologic (natural subsidence), biotic (flora, fauna and habitat) and human (in particular landscapes). The Lagoon is characterized by various landscape typologies due prevalently to the lagoon morphogenesis mostly resulting from the anthropic activity. In fact, it is possible to find at least three prevailing landscape typologies: those characterized by the fishing valleys (active valley landscapes); lakes with large stretches of water (valley and lake landscape) and those semi-natural (landscapes of the living lagoon), composed of deeper canals, "velme" (mud flats) and "barene" (shoals) punctuated with "ghebi"

[5]Plan for tourism in Jesolo Lagoon, ***2015***.

(small channels). Moreover, the three landscapes are sensitive from a natural and hydraulic viewpoint, which can be promoted for a controlled tourist use.

Consequently, in terms of methodology, the three landscape typologies may develop as many types of tourism with different intensity and forms, also in relation to the objects to insert (piers, docks, paths, viewing towers, stopover platforms, cycling and pedestrian networks, reconstruction or construction of fishermen huts, etc.).

The document has been laid down by analyzing different tools for the territorial and urban planning and programming, especially the UNESCO Management Plan "Venice and its Lagoon", and the Management Plan of the SCI/SPA Venice Lagoon, which have made it possible to identify precise objectives for a development of sustainable tourism in the Lagoon of Venice and consequently in the Jesolo lagoon (Fig. 2). Moreover, the on-site survey has been fundamental for understanding the part of the lagoon in the municipality of Jesolo; through the inspections, it has been possible to organize an album of the most significant landscapes so that they may be used as a fundamental part of the Strategic Master Plan. For a better management of the actions that could be carried out in the Lagoon of Jesolo, some examples of rehabilitation and use of this geographic scope have been presented (they may be realized also thanks to the EU funds shared with the Veneto Region) and some project

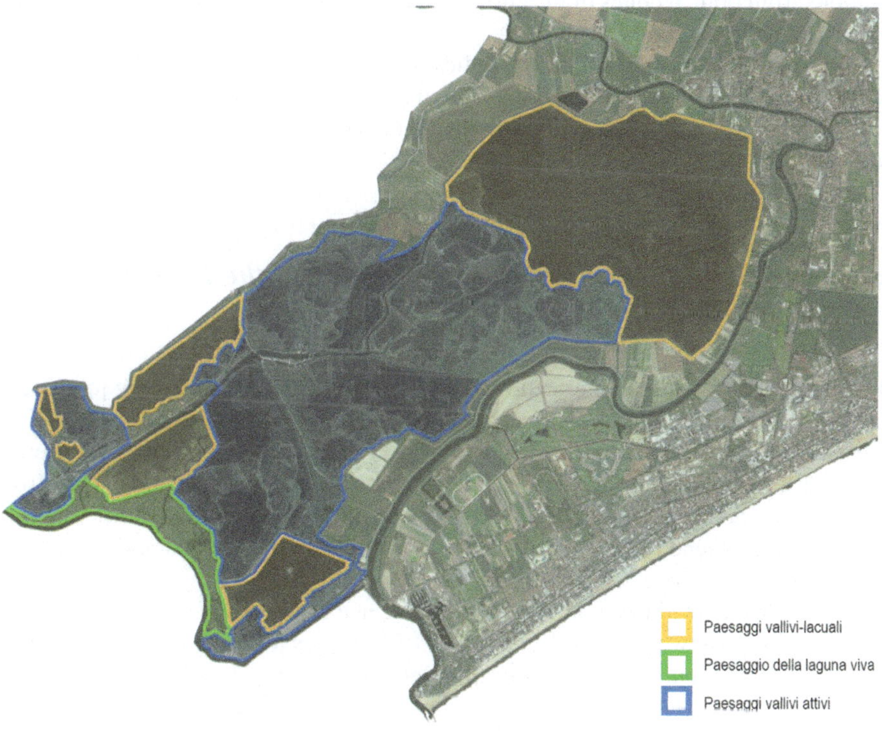

Fig. 2 Landscape of Jesolo Lagoon

abacuses have been elaborated for helping to plan actions for the tourist use, which will be accurately and punctually identified in the next Action Plan.

In detail, the work stages have been the following:

- territorial classification;
- elaboration of a planning and programming reference framework and relevant considerations;
- list of the restrictions existing in the Lagoon of Jesolo (Province of Venice);
- description of the lagoon of Jesolo and analysis of the existing naturalistic values and of the fishing valleys;
- list of the current initiatives in the lagoon and references to both national and international projects;
- evaluation of the lagoon landscapes;
- definition of the objectives of sustainability;
- identification of the significant landscapes;
- identification of projects and abacuses;
- connection with the PAT and governance (suggestions for the realization of actions through EU funding).

In each scope of the lagoon landscape, the types of action compatible with the features of the landscape and the tourist potential are indicated by means of photographic simulations from optical cones of the current and future states, of which some examples are presented (Figs. 3, 4, 5 and 6).

Thanks to these two specialized studies the municipality of Jesolo has an important wealth of helpful information for laying down the PAT, implementing it and establishing program agreements with the owners of the different parts of the Jesolo lagoon.

2.3 *The Naturalistic and Landscape Value of the Lagoon of Venice and Its Use*

The lagoon scope is characterized by extraordinary—and somehow unique—naturalistic, environmental and landscape values whose key element is represented by the anthropic and naturalistic presence. The lagoon is characterized by different types of landscape due prevalently to the lagoon morphogenesis and results from the complex anthropic activity that started with the foundation of Venice and was strongly strengthened after the determination of the lagoon boundaries[6] in 1791 by the Republic of Venice.

In fact, due to the human intervention the Venice Lagoon was modified in its natural dynamics. Between the XV and the XVIII centuries the Venetians of the Repubblica La Serenissima carried out imposing interventions for deviating the mouths of

[6]Characterized by 99 terracotta stones, which were replaced between 1846 and 1898 by as many, more resistant, Istrian stones.

Fig. 3 Simulation A

Fig. 4 Simulation B

Fig. 5 Simulation C

the main rivers that drained into the lagoon in order to prevent the Lagoon itself from filling up and to keep the peculiarity of this city as an island in the middle of water, all stretched towards the sea for trading and simultaneously protected against enemy raids from the inland.

The scope of the Venice Lagoon is characterized by three prevailing types of landscape: those characterized by the fishing valleys (active valley landscapes); lakes with large stretches of water (valley and lake landscapes) and those semi-natural (landscapes of the living lagoon), composed of deeper canals, "velme" (mud flats) and

to the conservation of these habitats and species, called Nature 2000 Sites and differentiated in Sites of Community Importance (SCI) and Special Protection Areas (SPA).

The "Habitats" Directive and the relevant implementing Presidential Decree 357/1997 provides for the realization of the Assessment of Implications on European Sites with a view to verify that the plans, projects or interventions, inside or outside the Nature 2000 sites, can determine significant negative effects (implications) in relation to the pursuit of conservation objectives for which these sites have been identified.

The municipal territory of Jesolo comprises three Nature 2000 sites: the Special Protection Area (SPA) IT3250046 "Lagoon of Venice", the Site of Community Importance (SCI) IT3250031 "Upper Lagoon of Venice", the Site of Community Importance (SCI) IT3250013 "Laguna del Mort e Pinete di Eraclea". For these sites a Study of Implications on European Sites has been laid down to verify whether the PAT of Jesolo has significant negative impacts on the Nature 2000 Sites involved, whose results have been integrated into the SEA for evaluating the effects on their biodiversity.

3 Results of the SEA Experience and Critical Reflections

The SEA process mainly operates through the elaboration of the Environmental Report, an analytical evaluation paper that shall contain all the analytical information and select the most suitable assessment models in order to verify, for the specific reality being examined, the level of interference of the planning tool with the more general reference environmental system. The environmental assessment is a tool that can credibly improve the plans and projects, but only under certain conditions, i.e. when the phase of environmental pre-feasibility is increasingly developed, when there is a virtuous synchronic (EIA) and asynchronous (SEA) collaboration between evaluator and planner, and the Public Administration contributes to the evaluation (methods, techniques, criteria, estimate of the impacts).

The SEA process is not produced to prevent the transformations, rather to improve the content of the plans and projects; its function is to support the strategy established by a local community; in fact, the environmental assessment is always a help to decision-making, which cannot replace the democratic decisions being taken by the Municipal Council. Therefore, the local community members can freely and lawfully decide on their future, consistently with the guidelines of the environmental assessment. In this sense, the Municipality of Jesolo could paradoxically decide to transform its tourist facilities into industrial/commercial ones, but also this choice should be made considering the environmental issues of the related evaluations.

The long road in the elaboration of the PAT of Jesolo has been supported by important in-depths analyses and evaluations leading to a plan that is not so "environmentally sustainable", but rather "environmentally performing", since it has tried to increase the social and economic opportunities through a clever "use" of the existing environmental values. In fact, there are at least two general transformation modes in the territory: conservative and evolutionary. With the former, the local community interprets the environmental issues as limits to "stop" the transformation processes; with the latter, the environmental issues are instead interpreted as "opportunities" for a high-quality development.

The experiences of the environmental assessment of plans and projects should make it possible to reach a cultural strategic objective, i.e. a "know-how" that may internalize the process of environmental assessment into the process of territorial transformation.

The evaluation carried out in Jesolo has also highlighted two important elements deserving attention: on one side, the importance of enjoying again the Lagoon of Venice, using the paper "Master Plan on Tourism in the Jesolo Lagoon"; on the other, the need to go beyond the limit of the landscape constraint imposed on the built-up coastal strip, which has largely lost the original meaning due to the transformations occurred in the meantime. Therefore, this constraint should be reinterpreted in the light of the paper called "In-depth analysis of the landscape evaluation of the urban and regional planning (PAT). Memorandum of understanding. Implementation of the PAT Overall Framework for the landscape art. 143, paragraph 1, Legislative Decree 42/2004, L. 16 April 1973, No. 171".

Consequently, with this analytical method the case of the PAT in Jesolo is an important example of methodological integration of specialized studies and evaluations associated with the structure of the municipal urban planning and the elaboration of the Implementing Technical Standards. In this way, it is possible to be consistent with the basic principles of the European Directive on the SEA that explicitly, at point 4) of the Preliminary Provisions, recalls the principle of environmental integration. In fact, it states that "*Environmental assessment is an important tool for integrating environmental considerations into the preparation and adoption of certain plans and programs which are likely to have significant effects on the environment in the Member States, because it ensures that such effects of implementing plans and programs are taken into account during their preparation and before their adoption*".

References

AA. VV (1989) Misurare la terra: centuriazione e coloni nel mondo romano. Il caso veneto. Modena

AA. VV (1992) La centuriazione romana tra Sile e Piave nel suo contesto fisiografico. Padova

AA. VV (2000) Il Piave (a cura di) Bondesan A, Caniato G, Vallerani F, Zanetti M Cierre. Verona

Busca A, Campeol G (a cura di) (2002) La valutazione ambientale strategica e la nuova direttiva comunitaria. Palombi Editore, Roma

Campeol G (1996) La valutazione ambientale nella pianificazione territoriale e urbanistica. In: Stanghellini S (ed) Valutazione e processi di piano (a cura di). INU-DAEST. Alinea Editrice, Campi (FI)

Campeol G (2000) La VAS del Piano Strategico della Provincia di Belluno. In: Governance, Pianificazione e Valutazione Strategica. Sviluppo sostenibile e governance nella pianificazione urbanistica (a cura di). Corrado Zoppi. Gangemi Editore, Roma

Campeol G. (2003) Un modello applicativo di valutazione ambientale strategica per i piani urbanistici, in Valutazione Ambientale. n° 3 Gennaio—Giugno 2003. EdicomEdizioni, Milano

Campeol G (2006) La valutazione ambientale dei Progetti e dei Piani. In: La riqualificazione della città e dei territori. Architettura e scienze a confronto (a cura di Fulvio Zezza), Quaderno Iuav 48 Dicembre 2006. Il Poligrafo, Padova

Campeol G (2007) La valutazione del paesaggio: aspetti metodologici e tecniche applicative. Ri-Vista ricerche per la progettazione del paesaggio n° 7 gennaio-giugno 2007. University Press, Firenze

Campeol G, Carollo S (2003) Modelli di valutazione ambientale per gli strumenti di pianificazione urbanistica: dagli indicatori ecologici a quelli paesaggistici. M, Zoppi C (ed) La valutazione ambientale strategica nella pianificazione territoriale (a cura di). Gangemi Editore, Roma

Campeol G, Carollo S (2004) Sviluppo sostenibile ed ecologia. Applicazione dei principi dello sviluppo sostenibile alla pianificazione territoriale e urbanistica. Individuazione di modelli per il calcolo della sostenibilità tramite indicatori. In: Atti del Convegno "Semplificazioni procedurali e operatività locale della nuova legge urbanistica della Regione Emilia Romagna". Federazione Ordini Architetti Emilia Romagna e Comune di Rimini, Rimini

Campeol G, Carollo S (2004) Sviluppo sostenibile ed ecologia. Applicazione dei principi dello sviluppo sostenibile alla pianificazione territoriale e urbanistica. Individuazione di modelli per il calcolo della sostenibilità tramite indicatori. In: Atti del Convegno "Semplificazioni procedurali e operatività locale della nuova legge urbanistica della Regione Emilia Romagna". Federazione Ordini Architetti Pianificatori Paesaggisti e Conservatori Emilia Romagna e Comune di Rimini, Rimini

Campeol G, Carollo S (2006) La VAS del PSC di Ferrara. In: Urbanistica Dossier n 88 supplemento al n 208 di "Urbanistica Informazioni" luglio-agosto 2006. Roma

Campeol G, Carollo S, Pypaert P (2002) Les principes de la durabilité à la planification des villes e du territoire: évaluation des modèles de plan de masse pour le villes côtières de Jesolo (Italie) et Omišalj (l'Île de Krk, Croatie). Actes du Séminaire international "Petites Villes Côtières Historiques: Développement urbain équilibré entre terre, mer et société". Saida Liban, 28–31 mai 2001, UNESCO, Paris

Campeol G, Črnjar M (a cura di) (2001) Regional planning and pilot projects for sustainable development in Croatia. UNESCO, Rijeka Hrvatska

Carollo S, Marcon R (1999) Tesi di laurea (relatore Giovanni Campeol) La valutazione della sostenibilità nei processi di trasformazione territoriale: il caso del Master Plan di Jesolo. Università Iuav di Venezia

Dorigo W (1994) Venezia sepolte sotto la terra del Piave. Jesolo

An Elaboration of the Land Use Plan in Relation to the Environmental Constraints. Social Demands

Renato Segatto and Daniela Vitale

Abstract The paper deals with the main stages in the elaboration of the Land Use Plan (PAT) of Jesolo, in compliance with Law 11/2004 of the Veneto Region. In this context, public participation is crucial since it enables to identify the planning matters that are strategically important for the preparation of the SEA.

Keywords Urban planning · Public participation · Legislation · Sustainability

1 Participation in the Urban Planning Law of Veneto

Since 23 April 2004, with Regional Law n. 11, the system of municipal urban planning has been deeply innovated. The previous and better-known General Urban Development Plan (PRG) has been divided in two different procedures called Land Use Plan (PAT) and Intervention Plan (IP). The former contains the planning strategic choices, recognizing at the same time the invariants and the indispensable territorial characteristics to be protected and promoted with the future territorial choices. The latter is more operational and analyzes in detail the various actions that can be performed in the municipal area.

The legislative innovation has also introduced mandatory information sharing with citizens about the elaboration of the tools prefiguring the future of Jesolo. In fact, Art. 5 titled "Concertation and Participation" of the Italian Law states that:

1. "In elaborating the territorial and urban planning tools the Municipalities and the Region shall develop their activities through debate and concentration with the other local public authorities and the other administrations responsible for protecting the public interests.

R. Segatto (✉) · D. Vitale
Public Works and Urban Planning Area, Municipality of Jesolo, Jesolo, Italy
e-mail: renato.segatto@comune.jesolo.ve.it

D. Vitale
e-mail: daniela.vitale@comune.jesolo.ve.it

G. Campeol (ed.), *Strategic Environmental Assessment and Urban Planning*, Green Energy and Technology, https://doi.org/10.1007/978-3-030-46180-5_10

2. The previous administration shall also ensure the debate with the economic and social associations that have important territorial or general interests, as well as with public service operators inviting them to contribute to the definition of the goals and strategic choices identified by the planning tools."

The initial activities for involving the community of Jesolo in the review procedure of the municipal general urban tools were inspired by the experiences already made while elaborating the present PRG. In that context, following the elaboration of the Master Plan by Kenzo Tange Associates of Tokyo, the Municipality of Jesolo laid down the set of documents that constitute the PRG. On that occasion, tens of meetings were proposed and organized with the aim to collect and share the guidelines to be pursued with the new urban tool, announcing what would then become a legal requirement.

Even then, the involvement of the population showed the big importance of this activity for at least two fundamental reasons: the former, and the most evident, is the one which makes it possible to collect the most complete information on the existing situation of the territory and to give the closest answers to the community's needs. The latter, perhaps less evident although fundamental, is the start—during these meetings—of a pathway for realizing a shared view on the future of the city. This aspect, which is surely measurable in the medium and long term, is essential for a successful planning both in more real terms as regards the realization of works, and in the collective imagination relating to the recognition of the city image, also beyond national borders.

Therefore, the participation already developed in the laying out of the first PRG has proved crucial not only in giving answers to the citizens but also in making all the community feel part of the choices, thus becoming more united and better characterized in front of a national and international public.

In this stage, a special attention has been given to all stakeholders operating in the municipal territory, starting from the public institutions, the representatives of the various professions, as far as infrastructure managers, service providers, sports associations, associations for environmental and animal protection, and any other spontaneous agencies representing the various areas of Jesolo territory.

All of them have been proposed special meetings in the town halls for enabling them to know the issue of discussion and the role to play, in different discussion opportunities, both individually and in homogeneous categories of stakeholders. The first plenary meeting in the town hall on 21 January 2011 was attended by a total of 123 associations out of the 147 invited.

All these operators were invited to the Council Chamber of the Municipality on 21 January 2011 with the Head of the Planning Department, several municipal employees charged with the elaboration of the PAT and the advisors responsible for the expert reports for the PAT. On that occasion, each participant was given the material elaborated until then: the Preliminary Document, the Environmental Report and the opinion expressed by the regional Committee for the Strategic Environmental Assessment.

The meeting was helpful to open the concertation-participation, to briefly illustrate the pathway for elaborating the PAT on the basis of the regional legislation, and to ask for the collaboration of the participants that should provide the Administrations with all the contributions of each specific reality so that the new planning might be closer to the real needs. After this, 18 meetings were organized with the most representative associations, during which the interventions were recorded.

Later on, one or more meetings were promoted with the single operator involved, in front of the Mayor, the Municipal Councilor responsible for the PAT, the Head of the Planning Department and a scribe. On these occasions, it was possible to analyze the matter, promote reciprocal reflection and collect pertinent and qualified suggestions. Moreover, most interlocutors had a written paper, a short report containing the contribution elaborated. It is thought that this form of participation was an effective witness of the best practices used by the municipal administration and of the high-quality answers obtained.

Beyond the meeting with the representatives of professions and associations, the Administration was interested in directly listening to individual citizens, in crowded places such as the 7 parish churches of the municipal territory and the weekly open-air market in the historic city center. In both cases, a Listening Point was organized around a gazebo and a camper van, operated by the Municipality staff, where citizens could not only ask questions but also leave post-it, requests and suggestions for making Jesolo become a city with an improving quality of life. More than 300 suggestions were collected, for which a summary was drawn up in categories, so that the post-it could be used to orientate the activities for the elaboration of the PAT.

The analysis of the material collected highlighted the quality, seriousness and importance of the suggestions received, which impressed for the rich proposals since they showed a remarkable sensitivity and deep knowledge of the territory expressed by the citizens involved, as well as a great variety of wishes.

In order to transfer all the material into documents ready for use and reading, it was necessary to list each contribution in homogeneous and significant categories, with the aim of definitely outlining the guidelines for each of the issues defined on the basis of the contribution collected, in view of future planning.

Later on, on the basis of this material, the relevant categories were established: each suggestion was ascribed to each category, deciding (for example) to build them on the effective basis of the citizens' contribution rather than theoretically pre-package them at the start of the activity.

More precisely, the following 7 categories were defined: mobility, urban spaces (places and buildings), services to citizens, healthcare, leisure and tourism, and seaside.

This paper reports below a synthetic summary of the material collected, as well as a more operational translation into guidelines for a concrete application in the PAT elaboration.

The participatory stage was concluded on 20 March 2012.

- **Mobility**
  - Improvement of cyclability and pedestrianization
  - Organization of pathways and access
  - Investment in sustainable mobility
  - Public services for citizens and tourists
  - Safety and quality of street furniture
  - Interconnection of urban mobility and access infrastructures.
- **Urban Spaces**
  - Eco-friendly buildings
  - Residential areas to be ensured in the city and in rural contexts
  - Organization of the system of green public areas
  - Definition of a big green park
  - Equipping the pinewood with compatible tourist services
  - Urban renewal and regeneration
  - Structures and services for new residents.
- **Services to citizens**
  - New services and socializing places for the youth
  - Centers for cultural services, entertainment and wellness
  - Organization of cultural itineraries
  - Renewal of public services and structures.
- **Healthcare and wellness**
  - Interventions in Social Housing
  - Renewal of the hospital
  - Investing in wellness facilities
  - Services for the elderly.
- **Environment**
  - Promotion of natural and landscape resources
  - Support to the sustainable use of the lagoon, valleys and rural landscape
  - Defense of the soil and rural territory
  - Prevention of water risks
  - Protection of the sandy beach.
- **Leisure**
  - Investing in sport, leisure and wellness
  - Facilities for entertainment, exhibitions and museums
  - Identification of a service center
  - Investing in sport facilities and structures

   - Improved offer of complementary experiences relating to the bathing resort sector.

- **Tourism and seaside**
   - Consolidation of the bathing resort sector with complementary products (yachting, nature, wellness, sport)
   - Better balance between seasonal trends
   - Development of hotels and open-air accommodation
   - Stronger services for tourists in natural areas and in rivers
   - New pathways and centers for complementary use
   - Rehabilitation, renewal and enlargement of accommodation facilities.

1.1 Analysis of the Contribution from Participation and Concertation

In front of such information it is interesting to propose here an analysis of the outcomes of the concertation and participation meetings, aimed at highlighting the most relevant and transversal information gathered in the different "tables".

Catchwords

After collecting the minutes of the Plan consultation steps, it has been chosen to carry out a study for analyzing this information in a view to highlight and transfer into the PAT the most important suggestions expressed by the different citizens and stakeholders.

The approach to the study is mediated by the semantic analysis: in fact, the key words used more frequently during the works of the different categories have been extracted from the minutes and then commented.

The analysis of the data has identified those key words that the city stakeholders have mainly underlined. Here below is shown a table with the most frequent words (Table 1).

The first analysis of the outcomes underlines the important role of recognisable socializing places such as squares and the city centre. A request is emerging to work on these places, which could strengthen the identity of Jesolo historic centre and the Lido.

The issue of residence as well as hotel and accommodation facilities is very important since, on one side, it is referred to living in Jesolo; on the other, it is reflected on the re-organization of hotels by identifying strategies and actions aimed to qualify the accommodation hospitality.

Sport is a particularly transversal word mentioned by various citizens, whereas the issue of parking lots is highlighted by specific stakeholders, like tradespeople and technicians; however, they both frequently appear. Nearly tenth in the ranking are words associated with the environment and the use of the territory: river, park,

Table 1 Most frequent words

Catchwords	Number
Main square	45
Centre	41
Residence	26
Hotels	25
Sport	23
Territory	23
Parking Lots	20
Citizens	19
Rooms	16
Sea	16
Commercials	15
River	15
Park	15
Pinewood	15
Cycling paths	14
Hamlets	14
Lido	12
Environmental	11
Services	11
Tourism	11
Buildings	10
Beach	10
Road System	10
Archaeology	9
Events	9

pinewood and cycling paths incidentally have the same number and they all go together with a request for the use of the territory and the development of tourist itineraries.

Hamlets, beaches, roads, events and the archaeological area are the last "hot" concerns that close the list of the most recurrent words, on which the Plan should focus its attention, and that should be considered in the PAT strategies (Table 2).

The further analysis of the key words subdivided by the various tables of concertation and participation leads to a deeper study of the matters. While confirming the direct interest for each topic in relation to the different stakeholders, the term hotels, for example, is mainly proposed by sector operators, but there are also common issues of general interest.

The square and the centre represent a clear example of transversal issues involving both associations and stakeholders, therefore the attention to these issues shall be

Table 2 Keywords from the participation

WORD	AJA	APT	ASCOM	AA_COMB	ASS_CULT	ASS_SPOR	ASS_CAT	COMITAT	COM_CENT	COM_PINET	COMM_SA	CONF_COM	ENTI	FORZ_ORD	ORD_PROF	PARR OCC	TECH	TOTALE
alberghi	8	7	2			1				2	2					1	2	25
ambientale							3	1					6		1			11
archeologica					5								1		3			9
camere	5	11																16
centro	2	3	5		3			5	15		1		1			6		41
ciclabili		2	3		3	3		3										14
cittadini				2	2		2	5						2		6		19
commerciali			6						7								2	15
edifici								2	3			1					4	10
eventi						4			1								4	9
fiume				3			2		5				2		3			15
frazioni					1	2		3								6	2	14
Lido					3			4									5	12
mare		3	2							4		2			2		3	16
parcheggi			6						6		4						4	20
parco									6	5	2				2			15
Piazza		6	5					7		7				5		15		45
pineta								11		2			1		1			15
Residenza							2	9	2	1	2	2	3			5		26
servizi	3	5					3											11
sport						10				7						3	3	23
spiaggia						3											7	10
territorio					4		3	3					1	7		2	3	23
turismo		2	2		2	1	1	3										11
viabilit‡							4							3			3	10
TOTALE	18	39	31	5	23	24	20	56	45	28	11	5	15	17	12	44	42	435

shared as much as possible in order to avoid any conflicts, but it also becomes a strategic element on which the Plan shall focus in order to effectively meet the requests reported.

Among the highest number of key words there are surely the committees, the parish churches—that have a high number—, and the same can be said about technicians and tourist operators.

A low number is referred to the requests of operators such as craftsmen; in fact, the most frequent word reported by Confcommercio is crafts but the low number does not put it in the ranking of the 25 most cited words.

The goal of this stage is to highlight some key elements that the PAT shall support and govern so that the concertation process may be recognized also within the strategic actions of the Plan.

All these operators were convened in the Council Chamber of the Municipality on 21 January 2011, at 6.00 PM, in front of the Mayor, various Aldermen and municipal Councilors, the Head of the Planning Department, several municipal employees involved in the elaboration of the Land Use Plan, and the advisors charged with the expert reports for the PAT. On that occasion, each participant was given the material elaborated until then: the Preliminary Document, the Environmental Report and the opinion expressed by the regional Committee for the Strategic Environmental Assessment.

The meeting was helpful to start the concertation-participation, to briefly illustrate the method for elaborating the PAT on the basis of the regional legislation, and to ask for the collaboration of the participants, which should provide the Administrations with all the contributions of each specific reality so that the new planning might be closer to the real needs.

Later on, one or more meetings were promoted with the single operator involved, in the presence of the Mayor, the Municipal Councilor responsible for the PAT, the

Head of the Planning Department and a scribe. On these occasions, it was possible to analyze the matter, promote reciprocal reflection and collect pertinent and qualified suggestions. Moreover, most interlocutors had a written paper, a short report containing the contribution elaborated. It is thought that this form of participation was an effective witness of the best practices used by the municipal administration and of the high-quality answers obtained.

The city has become a complex place due to fragmented processes, complex relations, and the contradictions of an increasingly technological society; governing this complexity is becoming more and more difficult, beginning with the joint evolution of economy and the environment.

This difficulty is also due to the distribution of competences between the various institutional authorities endowed with different forms of power on the territory; sometimes these powers and competences are not exercised in a subsidiary and cooperative way.

One of the methods (but not the last) to overcome these difficulties is the production of knowledge, to be socialized with all the actors of the government of the city and the territory, including private citizens.

In fact, the availability of knowledge also simplifies the management of complexities, making the different actors aware of the consequences, therefore of responsibilities, of their behaviors in the space and so in the environment,

The availability of knowledge also makes it easier to fulfill obligations, e.g. in terms of energy saving and CO2 emissions. Contemporarily, it facilitates the consortium's management of the environmental issue, since the knowledge of the environment and the environmental consequences is at the basis of common actions, thus helping to reduce the management costs of the same environmental problem.

Therefore, it becomes a factor of the competition, although respecting the conditions imposed by increasingly stricter rules and regulations to the benefit of a better quality of the environment and of life, which means for all those who stay, work and live in the territory.

The awareness of a "limited quantity" of many natural resources, of the urgent need to reduce consumptions and greenhouse gas emissions is progressively pushing the communities to voluntarily adopt climate policies, i.e. to pursue projects and internal environmental policies aimed to reduce the consumption of energy and to modify lifestyles, optimizing the administration of the city.

Moreover, the territory is a "limited quantity", such as the individual components of the environment, and the "environmentally-friendly" choices should be part of the knowledge to be shared.

This research with the University IUAV is oriented to procedures and requirements that can be the basis for evaluating the energy-environmental sustainability of the territory, in line with Directive 2010/31/EU (energy performance of buildings), implemented under national Law n. 90/2013.

Furthermore, with some operational-scale proposals, it develops the issue of mitigation and territory adaptation to the effects of ongoing climate change, as expressed by the EEA (2012) on "Urban adaptation to climate change".

Also the Ministry of the Environment, Land and Sea Protection has identified a series of "minimum" environmental criteria for buildings (construction and renovation of buildings), now mandatory for public buildings pursuant Art. 34 of the Legislative Decree n. 50/2016 "Public Contracts Code".

This is a procedure being progressively reinforced in international and community experiences, concerning the introduction of specific environmental and social criteria into public contracts.

This trend is expected to be strengthened in the near future thanks to the growing awareness of the public opinion and Public Administrations on the possible effects on the quality of life caused by the non-compliance with rules and controls on the environmental and social conditions.

2 Meeting Citizens

The concertation stage with citizens has been developed focusing, as specified in the introduction, on actions and initiatives carried out through a unique, coordinate communication characterized by a new logo.

Also in this case, to facilitate the reading of contributions, it has been necessary to list each intervention in homogeneous and significant macro-categories.

Therefore, the citizens' post-its have been organized in the following categories: mobility, urban spaces (places and buildings), services to citizens, healthcare and wellness, environment, leisure, tourism and seaside, other categories.

Mobility

1. Cycling path (4)
2. Extension of the cycling path in via Tram towards San Donà
3. Cycling path from Passarella to Jesolo (3)
4. Via Trinchet: Cycling path (2), road paving, guard rail, lighting
5. Cycling path via San Marco
6. Cycling path in Ca' Pirami (2)
7. Cycling path in via Ca' Gamba Paese Lido (8)
8. Cycling path from Piazza Milano to Piazza Drago
9. Cycling path from Piazza Milano to Jesolo Paese
10. Cycling path in via Colombo (2)
11. More cycling paths: from Eraclea to Jesolo (4)
12. Cycling path from Cortellazzo to Jesolo Paese (5)
13. Cycling path from Oasi di Jesolo via Rusti to crossroad Coop
14. More maintenance of the cycling paths along the left side of the Sile river (via Cristo Re towards the locks)
15. Pedestrian and bicycle connection between the artisan area and the village, via Vivaldi

16. Via Massaua and via Chiesa Chiesa should be opened to bicycles
17. The territory should be planned on the basis of the services to citizens and mobility (cycling paths and public transport)
18. Via Posteselle: lighting (2), guard rail, cycling path (3), road resurfacing
19. Unbearable smell due to the rearing of pigs
20. Road safety (2)
21. Controls in via Tram (5)
22. No buses or trucks in via Tram (3)
23. Restoration and adaptation of via Argine San Marco
24. ATVO service in via Tram for high-school students
25. Few coaches for Venice
26. Bus line for San Donà with a high daily frequency
27. Underground
28. Road-marking (lines) is missing
29. Cleanup of the road slopes from garbage thrown by drivers (with the help of volunteers)
30. Overhanging branches in via San Marco need cutting
31. Scandalous traffic-light system in Ca' Fornera
32. Sidewalks in Ca' Fornera
33. Lighting in via Castellana and via Fornera
34. Access ramp via Roma right-side in front of the supermarket PRIX
35. Via Correr: rejection of the road closure (2)
36. Missing sidewalks (2)
37. Improvement of via Fornasotto (2)
38. The roads linking the highway with four beaches of the Adriatic Sea in Veneto Region already exist. They should be widened by some meters and/or rectified with some road turns near small-inhabited centers. Altering or spoiling reclaimed agricultural land with new roads means violating these natural areas. Memory goes back to the last decade of flooding, overflowing, banks collapsing or breaking, with related damage to the people, the firms and all the community.

    a. Alvisopoli—Bibione
    b. Portogruaro—Caorle—San Stino
    c. Noventa di Piave—Eraclea Mare
    d. Meolo Jesolo
    e. Airport—Jesolo

39. Roundabout Jesolo – softer access to Eraclea.
40. Warehouses behind Piazza Torino are too close to the beach
41. Change the access (in front of Via Fornasotto)
42. Damaged streets
43. More cleaning of sidewalks, especially behind the Famila supermarket, the narrow streets of Cascina del Mar and Maricel
44. Roads, connections with Cavallino Treporti (2)
45. Sidewalks via Gorizia n 10 (on the side of Condominio Luisiana)

46. Service stations distributing methane for vehicles
47. Cycling path via Roma Destra: from the bridge of Cavallino to the roundabout Picchi
48. In the area of Piazza Trieste the road system needs improving: more signals and traffic control
49. Free parking lots to go shopping at the lido (there are just 3 of them in the accesses to the lido), improving roads and then houses, protecting the pinewood (2)
50. More roads and maintenance of the existing ones (2)
51. Pedestrianizing Piazza Matteotti
52. Accessible sidewalks
53. Higher presence of the local police to handle the traffic
54. Via Tram: speed control and higher lighting
55. Ring road Jesolo Paese—Lido, to innovate traffic towards the pinewood thus increasing mobility in favor of Jesolo compared with Cavallino. Freeing the traffic of via Roma Destra
56. Pedestrian paths for the town
57. Road surfacing of via Sant'Antonio, alleys 2 and 3; already referred to Begamo; are alleys state-owned?
58. Increasing local public transport in summertime for the elderly.

Urban Spaces

1. Promoting eco-friendly buildings (renewable energies, etc.) (2)
2. Restoring the rural annexes for residential purposes beyond the current active catalogue, envisaging the possible house extensions (20%) with the possibility to add part of the annexes although detached from the main office
3. Shameful parceling, houses stuck together, impassable roads, e.g. via Melograni
4. Considering the increase in buildings, new trees should replace cut down ones
5. For Ca' Fornera the area for new buildings should be broadened
6. Fewer buildings, more green areas (15)
7. More green areas and green trees (3)
8. Stop cement in Jesolo (4)
9. Plant trees and more vegetation and public green (2)
10. Failing ERP (public housing) policy, to be re-founded on clear fixed rules, transparency and short times (2)
11. Via Torcello: lift parking restrictions on the right side of the street
12. Deterioration of the area behind Piazza Torino
13. Better maintenance and interventions in the pinewood (e.g. benches and lighting) (4)
14. In the pinewood, more services for the commuting tourists
15. Too many variants compared with the previous PRG
16. Residents' homes

17. For the 9th street crossing Via Ca' Gamba, review residential area (rather than rural)
18. Restoring old houses, no cement (3)
19. The area Faro via Delle Vigne needs to be changed (a dock and a garden with flowerbeds are planned since it is an area for tourists' frequent walks)
20. The great Japanese architect had many ideas but he left a lot of chaos (area between the two pinewood campsites)
21. Extra cheaper or free car parks (3)
22. The problem of car parks is linked to the construction of apartments. The new buildings must be accompanied by a sufficient number of streets and parking lots
23. The urban destination of the ex-technical office in front of the church tower needs clarifying
24. Via Toscanini: replace wells, surface the bumpy street with asphalt, solve the problem of flooding by repairing the sewage system
25. A big public green park (like San Giuliano) (2)
26. Houses for young people at more accessible prices. Facilitate the construction for the residents' children
27. Equipped playground, green space with a fountain (2)
28. Playgrounds also for small children (2)
29. Children's parks in the area of Piazza Trento
30. Playground in Piazza Trieste (2)
31. Improvement of the water-scooping machine, more trees
32. Regulations for the maintenance of private ditches
33. Re-populating the historic center (3).

Services to Citizens

1. More activities for high-school teenagers
2. Youth policy is missing (young people are abandoning Jesolo) (2)
3. The youths have no socializing places (4)
4. A civic and cultural youth center
5. More attention to small children
6. Municipal infant-toddler center (5)
7. A bigger library
8. Public baths in the marketplace area (2).

Healthcare and Wellness

1. Nursing home (12)
2. For the residents, projects of SOCIAL HOUSING, that are already being developed in some municipalities of the Lombardy Region, should be developed
3. Hospital of Jesolo (18)

4. Spas for thermal treatment (3).

Environment

1. No composting (7)
2. Safety of the landfill (4)
3. No landfill (2)
4. No more garbage.

Leisure

1. New public swimming pool (5)
2. Residents' discount for entry tickets to Aqualandia water park
3. Cinema (3)
4. Failing to fulfill the promise about the rugby field; the town cannot die
5. Equipped areas for dogs
6. Leisure place with billiards and bowling alleys (2)
7. Better maintenance of the soccer field near the Rodari Schools
8. Non-profit sport facilities
9. More initiatives and fun for young people (2)
10. A SKATEBOARD park for young people.

Tourism and Seaside

1. More free beaches (2)
2. The beach is saturated, too many new buildings without beach umbrellas
3. The city is not livable at weekends (2)
4. Promoting initiatives and activities in wintertime (also through twinnings with other Italian towns)
5. Poor cleaning on the foreshore
6. More restaurants and shops
7. Designers' shops
8. No visitors' tax
9. No big coaches or groups of visitors in via Bafile and extensions.

Other Categories

1. Poor attendance of the Mayor (3)
2. A new positive class open to all citizens
3. Place yourselves in front of the window and observe the villages outside
4. Meetings with the committees in the single areas of the community centers

5. Problem linked to asbestos
6. No nuclear power plants.

List of contributions

The list here below summarizes all the requests received from the citizens since 10 March 2005, starting date for the concertation in compliance with Town Council Resolution n. 68, which approved the framework of the "Preliminary Document". For more comprehensiveness, this list also includes some requests preceding the start of the concertation, and the unaccepted comments with the partial variant to the 2005 PRG, called "adjustment variant", that are listed at the end of the same list with their original number.

The contributions received by the Protocol Section are not relevant to the strategic content of the "Preliminary Document" of the Land Use Plan, which is not a regulatory tool of the single property; therefore these punctual contributions, with this in-depth level, are transferred to the Intervention Plan (IP) and shall be presented again when elaborating the IP.

Proposals can be divided in different categories following the type of request, classified on the basis of the various issues and summarized as follows:

- Insertion of C lots;
- Land-use change of homogeneous territorial areas;
- Planning/program agreements;
- Change of use, from hotels to residential buildings;
- Executive plans;
- B-type intervention record sheets concerning historical and testimonial buildings;
- Rural annexes.

The estimate of the volume requested, elaborated by the office and/or contained in the contributions, is equal to 3,800,000 m^3, well over the projections of the current PRG.

Moreover, the contributions of all the representatives of the categories and associations working in the municipal territory (shown in grey in the enclosed list) shall be considered relevant to the structural character of the PAT, which outlines the strategic choices of planning and development for the governance of the municipal territory.

A brief summary of the contributions elaborated so far has led to a set of strategies for the various systems that characterize the city and can lay the foundations for the future strategies of the PAT.

3 Strategies for Jesolo in the P.A.T. Systems

The history of transformations that have characterized Jesolo is about a city that has continuously been committed to become a reference point of the Veneto coastal system in Europe and all over the world.

The change of lifestyles, more concerned with the principles of sustainability, environmental and landscape quality, introduces a new approach and a different ethics to direct the territorial transformations towards a vision aimed at meeting the new needs that arise from this social change.

The PAT is an opportunity to assume these principles and enrich the governance of transformations with new values that the various actors shall relate to in order to build a project of a sustainable, smart, efficient, solidarity-based and intrusive territory, particularly open to the international context for an attractive city in terms of initiatives and investments.

For doing so, the Plan shall also have an "introspective look", more focused on all those elements of environmental and territorial quality of the municipality that, in this process of valorization and promotion, are the new values to be communicated and reflected outwards with a view to re-launching the destination towards 2020–2030.

Today Jesolo aspires to be a sustainable, efficient, hospitable and attractive city for users and residents, to be a place able to combine and systematize residential areas, tourism and work, by exploiting the request for use in order to contemporarily offer services for the territory thanks to the integration of environmental, historical and cultural excellence.

Through the Plan, the city of Jesolo has the opportunity to reorganize the present and future experiences for developing the territory, carrying out these actions inside a network of strategies and values that may permit to govern and direct future developments towards the goal of urban quality.

In this moment, it is a matter of summarizing the contributions elaborated so far by building a strategy grid for the various systems that characterize the city, therefore laying the basis for defining the PAT specific guidelines.

The environmental system

The Plan considers the importance of protecting the natural, environmental and landscape resources of the Municipality, supports the valorization of the lagoon, valley, rural and coastal excellences through operations of active maintenance and integration, promoting a sustainable use of the areas. The strategies to be pursued are aimed to protect the sandy shores from erosion phenomena by verifying the various possible options, the safeguard of the pinewood as a strategic place for "open air" tourism. The natural resources of the lagoon, river and valley environments represent the favorite places where complementary products of the bathing resort sector can be offered. Therefore, the Plan favors the identification of pathways and itineraries aimed to discover the inland peculiarities as well as the lagoon and water systems.

In this design, the rural areas have a multifunctional role since they integrate high-quality production and marketing, the enjoyment of the environment with recreational and didactic activities, accommodation facilities, and activities for leisure and slow time. It is a system of rural parks in a public use network, which promotes the realization of natural itineraries with cycling and pedestrian paths, the introduction and enhancing of tourist establishments, recreational areas, sports activities and services.

The Plan supports the promotion and marketing of local products, as well as more eco-friendly crops, by protecting and valorizing the natural, environmental and landscape heritage. It also favors hydraulic risk prevention and the protection of rural areas from erosion.

The tourist system

The PAT pursues the goal to support initiatives for the development of accommodation facilities by favoring the reorganization and regeneration of the structures on the basis of a new supply model that may better satisfy the new kinds of holiday-making. Today, the demand for holiday-making and use of leisure time corresponds to the request for a high-quality environment and landscape, and is characterized by shorter holidays spread out the whole year round in search for diverse experiences. If once there was the almost exclusive interest for bathing resorts and entertainment, today a "creative tourism" is developing and it focuses on experience and on a more active approach to places where tourists try to live and slowly contemplate the territory with a special attention to the quality of life/well-being.

In order to meet these needs through the Plan it is necessary to activate a set of measures oriented to the renewal and regeneration of the existing structures, with possible volumetric incentives for operations that may adapt and broaden the accommodation supply. In this sense, on one hand the goal is to renew the size and number of rooms, also starting a more flexible administrative management of files (database of rooms and special desk); on the other hand, there is the intention to favor the improvement of services with complementary activities such as wellness, fitness and entertainment.

The environmental integration and sustainable use of the territory shall be ensured through the synergy with the environmental system, identifying in the Plan the strategies to enrich the bathing resort sector with complementary products (yachting, nature, wellness, sport, agro-food products). The transversality of the tourist system in Jesolo becomes crucial in the renewal of the offer; the PAT identifies in the central area of the Paese-Lido connection the strategic place where it is necessary to promote the development of public use spaces and functional attractive structures for exhibitions, entertainment and cultural events.

Moreover, the Plan aims to promote and identify spaces dedicated to the new emerging proposals such as thematic events that have recently been inspiring the City (Miss Italia, Sport and Testimonial), also promoting the proposals concerning hyper-specialized medical tourism (treatment center and rehabilitation of professional athletes) able to internationally draw the attention on the destination.

The urban system

From an urban viewpoint, the studies and proposals analyzed so far consider the Lido/Paese connection as crucial for the future development of Jesolo. Assuming this objective, the PAT identifies in the so-called "Campana" the linking space between the two systems, equipping this area with all those services and infrastructures that will be functional both to tourists and to the entertainment for Jesolo residents.

The Plan guidelines for urban development are intended, always in a view of sustainable city, to construct efficient and reversible buildings, also promoting conversion projects and recovery plans. This will make it possible, where necessary and for more functional access and area connections (also in relation to the new infrastructural road networks), to aggregate or transfer volumes. The completion of the Detailed Plan called "Campana" permits to identify urban elements and structures not only to achieve the Lido-Paese connection but also, at the same time, to expand the offer for stable residents. Through the PAT, the PRG projections arising from the Masterplan vision are confirmed, and therefore enable a partial transfer of the planned volumes when the ongoing infrastructural reorganization requires their use, although without deviating from the comprehensive view of Tange's design.

Urban regeneration also goes through the reorganization of green areas and socializing places, the renewal of the hospital district and of the central area of Jesolo Paese in a Plan design that considers the "Campana" as the heart of the new Jesolo. In this view, the municipal villages become strategic for the maintenance of the inland natural and rural system; they preserve the city traditions and therefore it is important to invest in those economic sectors that promote the rediscovery of history. To ensure living in the villages the Plan identifies and promotes the creation of primary services, rationalizing the road system and favoring the improvement of environmental quality, the necessary conditions to improve the quality of life and of residential areas.

The infrastructural system

According to the Plan, the infrastructural system has a strategic role; the multiple stages of study, planning and debate concerning the territory over the years present criticalities in relation to this system. At the same time, these criticalities show the opportunity of a reorganization of the infrastructures, especially if coordinated to the more general reorganization of the urban space.

Considering Jesolo as a developing network node, the PAT assumes what is planned in the regional system, considering and organizing its road access accordingly. The highway of the sea represents the new road access to the city, therefore through the Plan it becomes strategic to focus the attention on the access nodes of this infrastructure that shall be considered as the new gateway of Jesolo: an arrival and exchange point where to promote intermodal transport, reorganize the interconnection with urban mobility by means of Park and Ride areas, also exploring alternative solutions to the car such as tramways connected to the SFMR (Regional Metropolitan Railway Service). Also, the coastal connection needs being reinforced through interventions that shall promote the connection with Eraclea, on one side, and by sea with Venice, on the other. In this design of regional network, it becomes relevant to verify the possibility of an airfield in relation to tourist needs.

At an urban level, the Plan encourages investment on sustainable mobility (bike sharing, car sharing, electric vehicles, etc.) exploiting the demand for tourist mobility to integrate functional services for the residents. In this sense, the organization of dedicated cycling paths and itineraries connecting the Lido-inland becomes crucial for offering the citizens sustainable and efficient modes of transport, and also for favoring and promoting the discovery of the territory to those who visit and stay in Jesolo. The discovery of the territory is also referred to the waterways as a network to reorganize by defining connection and docking points through the Plan.

In this intermodal and functional reorganization, the Plan pays a special attention to the system of car parks and accesses to the sea. It envisages a careful location of parking areas along the Lido coast, favoring the arrival and exchange with other means of transport, in a view to ensure more sustainable urban mobility and to relieve urban roads from car traffic.

The productive and commercial system

The goal of the Plan is to promote the development of crafts through procedures that may simplify the administrative management and regulate organically the definition of these areas through a dedicated information desk. The integration between crafts, trade and accommodation acquires a strategic role in the historic center, with a natural commercial vocation, identifying through new functions the opportunity of renewal and revitalization interventions in Jesolo Paese.

The Plan aims to support also the commercial development of the Lido by re-projecting the trade areas in line with the new accesses, and by building covered places and paths able to promote the attractiveness of the coastal urban area and to enable its use throughout the year.

The recovery and connection between Lido and Paese goes through the identification and enhancement of the historical and archaeological heritage. In fact, the Plan envisages creating routes and itineraries that may promote the discovery of these areas and may also become new opportunities for developing commercial activities and services.

The cultural system

The cultural dimension of Jesolo represents a particularly strategic element to help re-launching the destination and promoting the historical and archaeological heritage of the city.

The PAT considers the "central park" as an area where to invest in structures and services oriented to entertainment, exhibitions and museums, a central hub for attracting and contemporarily connecting with the network of museum itineraries in the territory. In fact, the Plan aims to give the opportunity to discover the historical and cultural excellences of Jesolo through the creation of a museum itinerary that shall systematize the existing and future museums with the wealth of the territory, such as the valleys, the archaeological area, the reclaimed landscape and the modern structures of architectural value. This is an itinerary ranging from archaeology, nature, rural culture and food excellence to memory museums and contemporary architecture.

In Jesolo, culture is also synonym of events that the city increasingly hosts with the aim to favor a de-personalization process still going on. The PAT is intended to promote this sector through the identification of dedicated areas and related services for investing in "specialized" hospitality connected with exceptional events and testimonials, by intercepting and evaluating the feasibility of proposals submitted by sectorial stakeholders.

References

Bettinotti M (a cura di) (1996) Kenzo Tange. Architettura e disegno urbano. Electa, Milano
Carollo S, Marcon R (1999) La valutazione dello sviluppo sostenibile nei processi di pianificazione territoriale: il caso del Master Plan di Jesolo. Tesi di laurea (relatore Prof. Campeol G.). Università Iuav di Venezia
Secchi B (1989) Un progetto per l'urbanistica. Einaudi, Torino
Urbani P (2000) Urbanistica consensuale. Bollati Boringhieri, Torino
Vitale D (1992) La politica di piano di una città a vocazione turistica: il Lido di Jesolo. Tesi di laurea (relatore Prof. Ciciotti E.), Università Iuav di Venezia

The manufacturer's authorised representative in the EU is Springer Nature Customer Service Centre GmbH, Europaplatz 3, 69115 Heidelberg, Germany. If you have any concerns regarding our products, please contact ProductSafety@springernature.com

Printed and bound by CPI Group (UK) Ltd, Croydon, CR0 4YY
30/09/2026
02225710-0006